EXTRATERRESTRIAL LIFE

UFOs

Jim Whiting

San Diego, CA

© 2012 ReferencePoint Press, Inc.
Printed in the United States

For more information, contact:
ReferencePoint Press, Inc.
PO Box 27779
San Diego, CA 92198
www.ReferencePointPress.com

ALL RIGHTS RESERVED.
No part of this work covered by the copyright hereon may be reproduced or used in any form or by any means—graphic, electronic, or mechanical, including photocopying, recording, taping, web distribution, or information storage retrieval systems—without the written permission of the publisher.

LIBRARY OF CONGRESS CATALOGING-IN-PUBLICATION DATA

Whiting, Jim, 1943-
 UFOs / by Jim Whiting.
   p. cm. — (Extraterrestrial life)
 Includes bibliographical references and index.
 ISBN-13: 978-1-60152-172-9 (hardback)
 ISBN-10: 1-60152-172-3 (hardback)
  1. Human-alien encounters. 2. Alien abduction—Case studies. 3. Unidentified flying objects.
 4. Outer space—Exploration. 5. Science fiction. I. Title.
  BF2050.W46 2012
  001.942—dc22
                                                                                                  2011007025

CONTENTS

INTRODUCTION
Strange and Otherworldly Objects   4

CHAPTER ONE
The Birth of Flying Saucers   8

CHAPTER TWO
A Long History of Sightings   20

CHAPTER THREE
A Small Town in New Mexico   33

CHAPTER FOUR
UFO Encounters Get Personal   46

CHAPTER FIVE
What Should We Believe?   58

SOURCE NOTES   71

FOR FURTHER EXPLORATION   75

INDEX   77

PICTURE CREDITS   80

ABOUT THE AUTHOR   80

INTRODUCTION

Strange and Otherworldly Objects

On the morning of October 12, 1492, the Taino Indians on the island of San Salvador in the Caribbean Sea may have rubbed their eyes in astonishment. Three strange objects they had never seen before emerged from the direction of the rising sun in the east and slowly headed toward them, seemingly pushed along by billowing white clouds. These objects were actually three wooden ships with white sails, under the command of Christopher Columbus.

From the standpoint of the Indians, these ships were UFOs: Unidentified Floating Objects. Their inhabitants were completely alien to anything the Taino had seen before. As Columbus and his men splashed ashore wearing heavy gleaming steel armor over their clothing, they appeared to have come from a completely different world. The Indians, by contrast, were naked.

> **DID YOU KNOW?**
>
> Christopher Columbus may have seen a real UFO. According to a personal narrative of his first voyage to America, the night before he landed he observed an unexplained light in the sky moving up and down.

The Taino were especially fascinated by the weapons the intruders carried. "[The Taino] neither carry nor know anything of arms, for I showed them swords, and they took them by the blade and cut themselves through ignorance,"[1] Columbus wrote.

Modern-Day UFOs

About four and a half centuries later, a different kind of UFO grabbed newspaper headlines in the United States. These were Unidentified Flying Objects. They were as strange and otherworldly to Americans as Columbus's ships and men had appeared to the Indians. And as had been the case with the Taino and Columbus's swords, an undercurrent of potential military applications was often connected to UFOs.

Just as Christopher Columbus's three wooden ships must have seemed unreal to the Taino people of San Salvador, a UFO hovering over a modern landscape is hard to fathom. Yet reports of strange and otherworldly flying objects continue to this day.

The emergence of these modern-day UFOs came as the culmination of a process that began at about the same time as Columbus's historic voyage. For almost all of recorded history, humans believed that Earth was the center of the universe. Starting in the early 1500s, this belief began to crumble. New and improved telescopes kept pushing the boundaries of the universe outward. It soon became apparent that it contained millions and even billions of stars. Earth was just one tiny speck of cosmic dust that revolved around an average-size star. Yet as the importance of Earth in the scheme of the universe steadily shrunk, interest in explaining the mysteries of the universe expanded just as steadily.

A Universe Full of Possibilities

Writers began taking advantage of this newfound interest. In 1865 French novelist Jules Verne wrote a novel about space travel called *From the Earth to the Moon*. In the book, a huge cannon shoots two men to a landing on the moon. The book is considered one of the very earliest works of science fiction.

British novelist H.G. Wells continued the theme of moon landings in his 1901 work *The First Men in the Moon*. Those "first men" find an advanced civilization already living there. Wells is more famous for his 1898 book *The War of the Worlds*. It introduced an ominous theme into science fiction: Mysterious beings from the planet Mars invade Earth.

The idea of space travel involving Mars gained more popularity in the works of Edgar Rice Burroughs in the twentieth century. Most famous as the creator of Tarzan, Burroughs also wrote many works of science fiction that centered on Mars.

The emergence of comic book (and later movie) heroes such as Buck Rogers and Flash Gordon brought futuristic travel to an even wider

> **DID YOU KNOW?**
>
> A few months before the 1938 *War of the Worlds* broadcast, some soldiers in the Spanish Civil War may have seen the real thing. They reported seeing a glowing oval object about 32 feet (10m) in diameter and shaped like two inverted plates.

audience. So when noted actor and director Orson Welles presented a very realistic radio adaptation of *War of the Worlds* in 1938, many people thought Martians had actually launched an invasion. Widespread panic took place near the site of the "landing" in New Jersey.

The belief that intelligent beings existed somewhere in the universe—and had an interest in Earth, perhaps even in conquering the planet—was like a large pile of kindling. All it needed was a spark to set it off. That spark came on a sunny early summer afternoon in 1947.

The Birth of Flying Saucers

As he taxied his small plane down the runway of the airport in the small town of Chehalis, Washington, at 2:00 on the afternoon of June 24, 1947, Kenneth Arnold had no idea that he was about to make history.

Arnold was 32 years old. He grew up in North Dakota, where he became an Eagle Scout and a high school all-state player in football. He also swam and dived, well enough to try out for the 1932 US Olympic team.

He moved to the Pacific Northwest in his early twenties and worked for a company that sold firefighting equipment. Soon he established his own company, the Great Western Fire Control Supply Company. His customers were scattered in rural areas in five western states. It was easier to fly to his various destinations than to drive, so he bought a plane. Also, Arnold was married and had two daughters. He and his family owned a home in Boise, Idaho. In other words, Arnold was the model of a solid, responsible citizen.

History in the Making

On that soon-to-be historic afternoon, Arnold's destination was Yakima, Washington. The skies were clear, and Arnold had no reason to suspect anything other than a smooth, uneventful flight. The only distraction would be watching for a Marine Corps transport plane that had crashed near Mount Rainier several months earlier. The wreckage had never been located.

Towering more than 14,000 feet (4,267m), Mt. Rainier is the highest peak in Washington and second highest in the continental United States. It is surrounded by dense forests, and the Marine Corps had offered a substantial reward to anyone who could find the missing plane.

Flying at 9,000 feet (2,743m), Arnold did not see any sign of the downed aircraft as he made several passes over the most likely crash sites. Just before 3:00 p.m., Arnold decided to break off the search and head for Yakima.

A Flash of Light

A sudden bright flash of light off to his left startled him. Fearing he was too close to another plane, he quickly scanned the skies around him. He was amazed to see nine strange objects at about the same altitude as his. They were approaching the mountain from the north at a high rate of speed, perhaps 20 miles (32km) away.

To make sure he was not just seeing reflections in his window, he rocked the plane from side to side. Then he lowered the window. The objects remained clearly visible. Soon they passed in front of the snow-capped mountain, and he could see them more clearly.

> **DID YOU KNOW?**
>
> On one day alone—July 5, 1947—flying saucer reports came in from Philadelphia, Pennsylvania; Akron, Ohio; Springfield, Illinois; Los Angeles and San Jose, California; Augusta, Maine; Rogers, Arkansas; Prince Edward Island, Canada; New Orleans, Louisiana; Port Huron, Michigan; Augusta, Georgia; Walter, Oklahoma; and Seattle, Washington.

He estimated their length at between 60 feet (18m) and 100 feet (30.5m), with curving fronts. They were flat-bottomed and slightly rounded on top. One was larger and shaped somewhat differently than the others. To his astonishment, the mysterious objects did not have tail sections, an essential part of aircraft design.

Arnold was also struck by the variations in their flight path. They did not stay in a tight formation. They rose and fell and twisted as they flew. The sun glinting off them as they rotated caused the flash that had originally attracted his attention.

Arnold had enough presence of mind to check his watch when they reached Mt. Rainier. Less than two minutes later the strange formation passed Mt. Adams—about 50 miles (80.5km) southeast of Mt. Rainier—and disappeared. Those two minutes would change history.

Support from Others

He had no way of knowing it, but he was not the only person to observe the strange objects. A miner named Fred Johnson was working near Mt. Adams and happened to look upward. In a letter dated nearly two months later, Johnson explains, "I saw the same flying objects at about the same time. Having a telescope with me at the time I can asure [assure] you they are real and noting [nothing] like them I ever saw before they did not pass verry [very] high over where I was standing at the time."[2]

Several other people in Washington and Oregon came forward to say that they had also seen strange objects in the sky at about the same time as Arnold. And sightings were not limited to the Northwest.

A carpenter working on a roof in Kansas City, Missouri, said, "There were nine of them, flying in a group with one a little to one side. They were flying so fast I barely had time to count them before they were gone."[3]

A man in Oklahoma City said that he thought he had seen the same flight as Arnold, but before reading about him in the newspaper, he did not want to say much because he thought people would ridicule him. "I know that boy up there (Arnold), really saw them. . . . I thought it only fair to back him up."[4]

What Could They Be?

As Arnold continued to Yakima, several possibilities went through his mind. Because of their high rate of speed, he thought the objects might be military jets. But Arnold knew that no known military jets were as fast as what he had just seen.

Rockets had also emerged in the waning days of World War II. But they followed a fixed up and down flight path. The flight path he had witnessed was not only almost horizontal but also kept jiggling up and down.

A more sinister explanation also seemed possible. As soon as the guns stopped firing at the end of World War II in 1945, a new conflict emerged. The Soviet Union had been allied with the United States to defeat Germany. The alliance had been uneasy and distrustful.

The two nations had different types of governments. Now they were engaged in the Cold War. Tensions remained high, but no open fighting had occurred. Could Arnold have seen a secret Soviet weapon? It seemed unlikely. The Soviets probably would not want to test something in American territory.

Greeted with Disbelief

With images of the strange objects and possible explanations still buzzing in his mind, Arnold landed in Yakima. He told the airport manager, a

En route to Yakima, Washington, in 1947, pilot Kenneth Arnold (center) spotted nine fast-moving, saucer-shaped objects flying at the same altitude as his plane. Beside him are pilots Emil J. Smith (left) and Ralph Stevens (right) who reported a similar sighting 10 days later.

The Martians Are Coming!

An indication that Americans were primed to believe in UFOs came the evening before Halloween in 1938. Famed actor/director Orson Welles and the actors in the Mercury Theatre troupe made a broadcast of English novelist H.G. Wells's *The War of the Worlds* on the CBS radio network. The book recounts a fictional invasion of Earth by Martians.

The hour-long broadcast took the format of an on-the-spot news report that originated near the "landing point" in New Jersey. At the top of the show, Welles announced that what listeners were about to hear was not real. Some people may not have been paying attention. Others tuned in late after listening to programs on other stations.

The presentation was very realistic. Because there were no commercial breaks, horror piled on horror as the invaders supposedly began to spread out over the country.

As a result, thousands of people in the eastern United States believed that a Martian invasion was actually taking place. Some fled into the streets in panic. Others grimly waited in their homes with their guns, ready to blast the invaders.

The novel was made into a film in 1953 and set in Southern California. Tom Cruise starred in a 2005 version.

good friend named Al Baxter, what he had seen. Baxter laughed. He did not believe Arnold. The story might have died there. But someone called Pendleton, Oregon, Arnold's next stop.

When Arnold arrived there, several pilot friends greeted him. They did not share Baxter's doubts. "Several former Army pilots informed me that they had been briefed before going into combat overseas that they might see objects of similar shape and design as I described and assured me that I was not dreaming or going crazy,"[5] Arnold said.

With some time on his hands, he performed the calculations to determine how fast the objects had been traveling. He was astonished to come up with the figure of 1,760 miles per hour (2,832kph). Fearing that no one would believe him, he lowered the number to 1,200 mph (1,931kph), still far faster than any known human-made aircraft.

The following morning Arnold visited the offices of the local newspaper. Reporters Bill Bequette and Nolan Skiff interviewed him and filed a quick story just before the newspaper's impending deadline. A key element was Arnold's description of the flight path. He told the reporters that it was like skipping a flat saucer across the water. The reporters included that phrase in the story.

Then they went to lunch. When they returned, Bequette said that "the receptionist's eyes were as big as saucers—the kind we use under coffee cups! She said newspapers from all around the country and Canada had been calling. They wanted more details on the 'flying saucers.'"[6]

> **DID YOU KNOW?**
> By the end of July 1947, there were more than 850 media reports of flying saucers. That is an average of about 23 each day.

"Flying Saucers" in the Headlines

Some of those newspapers had taken Arnold's phrase about skipping flat saucers across the water, changed it a little bit, and put it into their headlines. The era of "flying saucers" had begun.

Realizing he had a huge story on his hands, Bequette sat down for an in-depth interview with Arnold. He wrote a much longer article, and it appeared on the front pages of newspapers all across the country.

During this time, Bequette formed a clear opinion of his subject: "Mr. Arnold did not impress me then as a person who 'saw things.' And Nolan Skiff also believed Mr. Arnold to be an honest and sincere person who was genuinely puzzled by what he had seen that day."[7]

It did not take long for Arnold to realize how sensational his observations had become. Within hours, a preacher from Texas called him and said the objects revealed that the end of the world was about to occur.

While he was eating in a Pendleton café, a woman caught sight of him and burst out the door, shrieking "There's the man who saw the men from Mars. I have to do something for the children."[8] But Arnold had not hinted about the possible origin of the mysterious objects, and he certainly never mentioned anything about Mars.

Believable or "Bug-Eyed?"

That was just the tip of the iceberg. Arnold quickly realized that not everyone shared Bequette's opinion of his reliability. An Oklahoma newspaper, for example, referred to him as "bug-eyed." Some military officials and scientists suggested that he had been mistaken in his observations.

"This whole thing has gotten out of hand," Arnold said. "I want to talk to the FBI or someone. Half the people I see look at me as a combination Einstein, Flash Gordon, and screwball. I wonder what my wife back in Idaho thinks."[9]

By then increasing reports of "flying saucer" sightings were flooding the nation's media. One of the most dramatic came from United Airlines pilot Emil J. Smith, a pilot for more than 10 years. He flew military transport planes during World War II and joined United when the war was over. Because some of his customary routes lay in the same area as Arnold's sighting, newspapermen quickly asked for his opinion about what Arnold had supposedly seen.

> **DID YOU KNOW?**
>
> Some of Kenneth Arnold's pilot friends said he might have seen a new type of American jet on that day in 1947 when he spotted nine strange objects flying near his plane. But not even models being tested could fly anywhere near the speed he had observed.

"I've never seen anything like that (Arnold's flying saucers) and the boys (other pilots) say they have not either . . . what that other fellow (Arnold) probably saw was the reflection of his own instrument panel," Smith replied. "I'll believe in those discs when I see them."[10]

Birth of a Believer

Seeing became believing on the night of July 4. Soon after taking off from Boise on a cloudless evening, Smith and copilot Ralph Stevens saw five objects approaching them head-on. One was slightly larger than the others. Stevens thought they were aircraft and blinked his landing lights to warn them of a possible collision. The objects quickly swerved away, then flew parallel with the airplane. The two men called a flight attendant to join them. She confirmed the sightings.

The mysterious objects continued to shadow the aircraft for at least 10 minutes before turning away and disappearing. As the three crew members talked about what they had seen, four similar objects suddenly appeared. Then they too vanished into the night sky.

"The discs were flat and roundish," Smith and Stevens told reporters when they landed. "They definitely were not aircraft."[11]

The UAL flight crew was far from alone in saucer sightings that night. "The nation was baffled Saturday [the following morning] by the 'flying saucers' reported seen in 28 states by hundreds of persons while conjectures on their meaning flew as furiously as the reported speed of the silvery discs," wrote the *Portland Oregonian*. "Near unanimity was recorded on some of the discs' characteristics—terrific speed, bright reflections, round or oval in shape, flat, and flying with a peculiar undulating motion."[12]

Smith became an instant believer. He must have contacted Arnold soon after his flight. By this time, Arnold was becoming very frustrated that no one connected with the military or the government had contacted him. He sent a telegram to the commanding general at an important military airfield. The general had requested a written report from Arnold right after his flight. Arnold quickly sent a report to him but had not received a response.

> **DID YOU KNOW?**
>
> Harold Dahl claimed that UFOs he encountered near Maury Island dropped heavy lava-like rocks on his boat, killing his dog and damaging the vessel. Skeptics doubt Dahl's story, saying the rocks could have come from a nearby smelting operation.

Getting the Military Involved

Arnold's telegram, in part, read, "Capt. Smith and myself have compared our observations in as much detail as possible, and agreed we have

Becoming a Believer

On the morning of July 8, 1947—two weeks after Kenneth Arnold's UFO sighting—Lieutenant Joseph C. McHenry told fellow personnel at the Muroc (California) Army Air Force Field that "someone will have to show me one of these Disc[s] . . . before I will believe it." Moments later, McHenry stated in a previously confidential affidavit, that he "observed two (2) silver objects of either a spherical or disc-like shape, moving about three hundred (300) miles an hour, or perhaps less, a[t] . . . approximately eight thousand (8000) feet, heading at about three hundred twenty degrees (320) due north."

McHenry immediately summoned several other people, who confirmed the sighting and agreed that the objects seemed to be flying disks. They vanished, but another mysterious aircraft soon appeared and circled the airfield. "From my actual observance the object circled in too tight a circle and too severe a plane to be any aircraft that I know of," McHenry said. "It could not have been any type of bird because of the reflection that was created when the object reached certain altitudes. The object could not have been a local weather balloon for it is very impossible that a weather balloon would stay at the same altitude as long and circle in such a consistent . . . nature as did the above mentioned object."

Despite signed affidavits from McHenry and eight other people, the air force declined to investigate the sighting.

Quoted in Project 1947—Muroc UFO Incident 8 July 1947. www.project1947.com.

observed the same type of Aircraft as to size, shape and form. . . . It is to us of very serious concern, as we are as interested in the welfare of our country as you are."[13]

The message seems to have gotten a quick response. Given the speed with which the story was spreading, the military could not ignore it any longer. On July 7, the army air force's Air Intelligence Requirements Division decided to look into the situation. A few days later, two army air force officers—Lieutenant Frank M. Brown and Captain William Davidson—interviewed Arnold.

In his official report, Brown came to the same conclusions about Arnold as had the reporter Bequette:

> It is the personal opinion of the interviewer that Mr. Arnold actually saw what he stated that he saw. It is difficult to believe that a man of Mr. Arnold's character and apparent integrity would state that he saw objects and write up a report to the extent that he did if he did not see them. To go further, if Mr. Arnold can write a report of the character that he did while not having seen the objects that he claimed he saw, it is the opinion of the interviewer that Mr. Arnold is in the wrong business, that he should be writing Buck Rogers [a comic book and movie futuristic space explorer] fiction.[14]

Brown added that Arnold almost wished he had kept his mouth shut. "Mr. Arnold stated further that if he, at any time in the future, saw anything in [the] sky, to quote Mr. Arnold directly, 'If I saw a ten story building flying through the air I would never say a word about it,' due to the fact that he has been ridiculed by the press to such an extent that he is practically a moron in the eyes of the majority of the population of the United States."[15]

Men in Black

The Arnold interview was not Brown's only involvement with flying saucers. One of the most bizarre flying saucer stories during the furor in the summer of 1947 came almost literally in the shadow of Mt. Rainier. A man named Harold Dahl said he had been cruising in his boat on June 21—three days before Arnold's sighting—near Maury Island, from which the

> **DID YOU KNOW?**
>
> The day after Harold Dahl told a friend about his encounter with six UFOs, a man in a black suit came to his house to warn him against repeating the story. Dahl's "man in black" later became the inspiration for the humorous Men in Black films starring Tommy Lee Jones and Will Smith.

towering peak is clearly visible. Six doughnut-shaped objects about a hundred feet in diameter began hovering over him. At one point, one of the objects dumped some heavy lava-like rocks on his boat, doing some damage and killing his dog. When he got back to the dock where he kept his boat, he told a friend what had happened.

According to Dahl's account, a man dressed in a black suit and driving a black car came to his house the following day. The man "advised" him not to tell anyone about what happened the previous day. If Dahl did, said the man in black, his family might be harmed.

Dahl disobeyed the advice. In late July he met secretly in a hotel room with Arnold, Smith, Brown, and Davidson. He reportedly turned over some of the rock samples to the officers.

They loaded the samples into a B-25 bomber and headed for their base in California. Not long after takeoff, one of the engines burst into

Pilots who reported seeing UFOs in 1947 all described them in remarkably similar terms: They had flat bottoms and rounded tops and flew at unusually high rates of speed.

flames. Two crew members bailed out. Brown and Davidson remained at the controls. The plane crashed, and they were killed. Their commanding officer later stated that he did not know anything about any strange rocks they might have been carrying. And Dahl himself later said that the whole story was a hoax.

Some people do not accept these explanations. They say that the commanding officer's denial does not mean anything. The military and government often deny things later proven to be true. A few people even suggest that the plane may have been sabotaged to keep the rocks from being tested. According to some reports, soon after this crash Arnold's plane had engine trouble, too. Only his skill as a pilot prevented him from crashing. As to Dahl's claim of a hoax, these people say that he may simply have been protecting his family.

Whether Dahl's account is true or not, one thing is certain. Arnold's flight launched the era of flying saucers. The world would never be quite the same.

A Long History of Sightings

As journalists like to say, Kenneth Arnold's story had "legs." It did not go away either quietly or soon. Instead, the incident seems to have burst a dam, finally releasing pressure that had been building up for thousands of years. Ever since humans began keeping written records, some kinds of UFO sightings have been noted.

It is tempting to dismiss many of these sightings as celestial phenomena that have scientific explanations previous generations would not have known or understood. The problem with this idea is that mysterious sightings keep occurring. Some can eventually be explained, but others remain in the realm of mystery.

See Like an Egyptian

What seems to be the first recorded sighting came during the reign of the Egyptian Pharaoh Thutmose III (1504–1450 BC). According to an anonymous scribe,

> There was a circle of fire coming from the sky. It had no head. From its mouth came a breath that stank. One rod long was its body and a rod wide, and it was noiseless. . . . Now after some days had gone by, behold, these things became more numerous

in the skies than ever. They shone more than the brightness of the sun, and extended to the limits of the four supports of the heavens. Dominating in the sky was the station of these fire circles.[16]

Thutmose responded with a show of force: "The army of Pharoah looked on with him in their midst. Thereupon the fire circles ascended higher in the sky towards the south. Fishes and winged animals or birds fell down from the sky. A marvel never before known since the foundation of this land!"[17]

Around 1500 BC an anonymous scribe recorded a possible UFO sighting in ancient Egypt. According to the scribe, Pharaoh Thutmose III (pictured) witnessed brightly lit "fire circles" coming from the sky.

A rod is 16.5 feet (5m). Some commentators say that the relatively small size of the "fire circles" means they must have had a mother ship hovering nearby.

In the Bible

Several centuries later, the biblical prophet Ezekiel wrote,

> As I looked, and, behold, a whirlwind came out of the north, a great cloud, and a fire infolding itself, and a brightness was about it, and out of the midst thereof as the color of amber, out of the midst of the fire, also out of the midst thereof came the likeness of four living creatures. And this was their appearance; they had the likeness of a man. And every one had four faces, and every one had four wings. And their feet were straight feet; and the sole of their feet were like the sole of a calf's foot: and they sparkled like the color of burnished brass (Ezekiel 1:4–7).[18]

Josef F. Blumrich, who spent many years working in the aerospace industry, sees the entire chapter as a detailed description of an alien spacecraft. For example, he believes the feet mentioned in the passage are actually the base of landing legs similar to those on American lunar landers, complete with shock absorbers and round footpads.

He concludes, "The over-all result, then, is a space vehicle technically feasible beyond doubt and very well designed to suit function and purpose; its technology is in no way fantastic but, even in its extreme aspects, lies almost within our own capabilities of today. The results indicate, moreover, that Ezekiel's spacecraft operated in conjunction with a mother vessel orbiting the earth."[19]

Ancient and Medieval Sightings

The ancient Greeks and Romans also appear to have had their share of sightings.

According to one story, in 329 BC Alexander the Great was leading his army toward the Jaxartes River in Central Asia. Two flame-spitting silver shields dove toward them, and the crossing had to be postponed

UFOs in Medieval and Renaissance Art

Works of art created hundreds of years ago may provide some of the most interesting—and perhaps compelling—evidence for the existence of UFOs. *The Crucifixion*, a 1350 fresco in a monastery in Kosovo, may be the earliest illustration of UFOs. It depicts two small, flying, triangular-shaped objects. Each contains a seated figure that appears to be piloting the strange object.

Published in 1493, the *Nuremberg Chronicle*—an illustrated world history based on the Bible that is among the first printed books in existence—contains what might be another pictorial representation of a UFO. Referring to an incident first reported in 1034, the book shows a fiery sphere streaking across the sky.

Carlo Crivelli's *The Annunciation with St. Emidius*, painted half a century later, shows a saucerlike object in the sky emitting a beam of light through a solid wall toward the Virgin Mary as she kneels in prayer. Painted at about the same time, *Madonna col Bambino e San Giovannino* (attributed to several different artists) depicts Mary with Jesus and John the Baptist. A saucerlike object with numerous short rays emanating from it is behind her left shoulder. A man and dog in the background gaze at it.

Aert de Gelder's *The Baptism of Christ*, painted about 1710, shows what seems to be a flying saucer hovering above Mary and Jesus. Four beams emanating from the saucer drench them in bright light, with the onlookers in deep shadows.

until next day when the mysterious objects disappeared. But no direct evidence that this actually happened seems to exist.

Several respected Roman historians—including Cicero (106–43 BC), Livy (59 BC–AD 17), Pliny the Elder (AD 23–79), Dio Cassius (AD 155 or 163/164–after 229), and Julius Obsequens (mid-fourth century AD)—collectively recorded nearly 150 sightings. Typical was Obsequens's account: "Near Spoletium a gold-colored fireball rolled down to the ground, increased in size; seemed to move off the ground toward the east and was big enough to blot out the sun."[20]

Periodic sightings popped up around the globe in the centuries following the demise of the Roman Empire. A tenth-century translation of the Sanskrit text of the Prajnaparamita Sutra contains several illustrations that appear to be flying-saucer-like objects. One shows two round airborne objects with domes in the center. Merchants in Cairo, Egypt, reported noisy aircraft hovering overhead in 1027. On the eve of a battle in Japan in 1235, the leader of one of the armies saw a flurry of round bright lights swirling and swaying in the night sky. Scientists he summoned to explain the phenomenon had a simple answer: The wind was blowing the stars.

> **DID YOU KNOW?**
>
> Written at about the same time as the birth of Jesus, the ancient Indian epics Mahabharata and Ramayana both describe combat with airborne invaders flying disklike devices called *vimanas*.

By then, numerous UFO sightings had been recorded in Europe. England was a particular hotbed. The first sighting was in AD 497, with five more in a 33-year period from 1167 to 1200. Many were seen on the continent as well.

The United States became involved in UFO sightings in late 1896. Thousands of people in California reported seeing what became known as the "Great Airship." Sightings continued progressively eastward for six months. Accounts described the craft as cigar-shaped with tapering ends, with most witnesses placing its length at more than 100 feet (30.5m). The overall appearance was similar to zeppelin airships, which began undergoing tests in Germany in 1900 and were completely unknown in this country.

Foo Fighters

One of the more widely reported rashes of sightings came near the end of World War II in Germany. Allied pilots reported seeing balls of brightly colored lights about 10 feet (3m) in diameter. Some airmen believed they were secret German weapons. But they made no effort to attack, not even when jittery gunners fired at them. These mysterious objects became known as "foo fighters," which may have derived their name from a popular comic strip of the era, *Smokey Stover*. One of the title character's catchphrases was "Where there's foo [probably from the French word *feu* for fire] there's fire."[21]

According to a contemporary account in *Time* magazine, "Their descriptions of the apparition varied, but they agreed that the mysterious flares stuck close to their planes and appeared to follow them at high speed for miles. One pilot said that a foo-fighter, appearing as red balls off his wing tips, stuck with him until he dove at 360 miles an hour (579kmph); then the balls zoomed up into the sky."[22]

When the war ended, Allied intelligence officers scoured German files for evidence that foo fighters had been enemy weapons. None was found, and they learned that German pilots had reported similar sightings.

Starting in 1896, thousands of people in California and farther east reported a cigar-shaped craft flying overhead. By all accounts, the unidentified craft resembled a zeppelin airship (pictured)—a German vessel that had not yet become operational in Europe or the United States.

Further research revealed that foo fighters had even appeared in the Pacific theater of operations, with both American and Japanese pilots observing them. Ball lightning may account for some of the sightings. This is a somewhat mysterious atmospheric phenomenon in which glowing spheres up to 3 feet (1m) in diameter appear during thunderstorms and last for several minutes. However, no comprehensive explanation of the sightings has ever been issued.

The Military Takes Notice

In the wake of the furor generated by Arnold's sightings, the newly formed US Air Force (up to this point it had been part of the US Army) formally inaugurated a program called Project Sign in early 1948. It was located at Wright Field (now Wright-Patterson Air Force Base) in Ohio and used the services of the Air Technical Intelligence Center, whose responsibilities included gathering information on foreign aircraft. It was classified top secret. As one officer explained, the classification came about by accident: "The only reason for the original classification was that when the project first started the people on the project did not know what they were dealing with and, therefore, unknowingly put on this high classification."[23]

> **DID YOU KNOW?**
>
> Nearly every ancient Olympic event (such as boxing, wrestling, running, and the javelin throw) has military applications, but the discus throw does not. Some people maintain that the discus was modeled after a flying saucer.

At about the same time, the flying saucer phenomenon claimed its first victim. Kentucky police said they saw a large saucer. By chance, a flight of four P-51 Mustang fighter planes was in the vicinity. Mustangs were one of the premier aircraft of World War II and had a strong record of durability. The flight leader, Captain Thomas Mantell, made visual contact and radioed that he was trying to get a closer look. Soon afterward, his plane crashed. Mantell had not tried to bail out. Why he crashed was never determined.

Project Sign would eventually evaluate 243 reports. Its officers produced an "Estimate of the Situation," which included sightings by particularly trustworthy observers that could not be explained by natural phenomena. The estimate concluded that the sightings were of real objects capable of maneuvers far beyond those of any known aircraft and suggested that they were of extraterrestrial origin.

The report made its way to air force chief of staff Hoyt Vandenberg, who was outraged and claimed the evidence did not support such a radical conclusion. He ordered the immediate destruction of every copy.

Project Sign was terminated in early 1949. It had one lasting accomplishment: hiring eminent astronomer J. Allen Hynek. At the time Hynek was director of Ohio State University's McMillin Observatory. Hynek did an initial screening to weed out cases that ultimately could be traced back to astronomical phenomena such as planets, stars, and meteors. Initially skeptical, Hynek would eventually become a strong proponent of the existence of UFOs.

> **DID YOU KNOW?**
>
> What is perhaps the most famous celestial object of all time, the Star of Bethlehem, acts in accordance with many contemporary descriptions of UFOs. "It stopped over the place where the child was," according to the book of Matthew (Matthew 2:10).

The air force almost immediately replaced Project Sign with Project Grudge, though with one important difference. As writer Mark Pilkington notes, Project Grudge had "the express purpose of publicly playing down the flying saucer sightings and diminishing public enthusiasm for the subject. . . . To the Air Force's relief, their strategy seemed to work: by the end of the year it seemed that they might finally have washed their hands of flying saucers."[24]

During the life of the program, 23 percent of the sightings could not be explained. Nevertheless, Project Grudge administrators recommended cutting back on the scope and intensity of the investigations when the program ended at the end of 1949. Eventually it consisted of a single officer. With the national press now virtually ignoring flying saucers, no one really cared.

Rise and Fall of Project Blue Book

Yet the original reason for the existence of Project Sign and Project Grudge did not go away. The Air Technical Intelligence Center continued to collect sighting reports. It received an unintended boost with the outbreak of the Korean War in 1950, when many reservists were recalled to active duty. One was Captain Edward Ruppelt, who soon made flying saucers his area of interest. At his urging, official research was revived in 1952 under the name of Project Blue Book, with Ruppelt in command.

World War II pilots flying in Europe and the Pacific reported brightly colored balls of light that appeared to follow them at high speed for miles. The glowing spheres, known as foo fighters, are depicted in this computer illustration.

Ruppelt originated the phrase "unidentified flying object," or UFO. He felt it sounded more neutral than terms such as "flying saucer" and "flying disk." It was also a more accurate description of the phenomenon and did not have the extraterrestrial connotations that had quickly become attached to "flying saucers."

His timing was good. The number of UFO sightings took a huge jump that year. Some took place over the nation's capital in Washington, DC. As UFO researcher Bill Yenne notes, "These incidents, which took place on the nights of July 19–20 and July 26–27, represented the biggest concentration of UFOs in recorded history and convinced a great many people that *these* unidentified flying objects were real."[25]

Apparently those convinced did not include the air force, which regarded Project Blue Book more as an exercise in public relations than a genuine effort to find the truth about UFOs. Nevertheless, it continued to catalog reports and investigate many of them. One of the more notable cases came in 1964, when New Mexico patrol officer Lonnie Zamora was in hot pursuit of a speeding car. He broke off the chase when he heard a loud roar and spotted flames from the nearby desert and drove off-road to investigate. To his astonishment, he saw a small egg-shaped object with red markings. Even more astonishing, two 4-foot-high humanoids (1.2m) stood next to it. They scrambled back into the mysterious object and zoomed away. Zamora and another officer found burn marks and indentations on the ground.

According to Major Hector Quantanilla, the then-head of Project Blue Book,

> There is no doubt that Lonnie Zamora saw an object which left quite an impression on him. There is also no question about

> **DID YOU KNOW?**
> Astronomer J. Allen Hynek had a small role in the popular 1977 film *Close Encounters of the Third Kind*. Many scenes in the movie were based on actual accounts of encounters with aliens.

Zamora's reliability. He is a serious police officer, a pillar of his church, and a man well versed in recognizing airborne vehicles in his area. He is puzzled by what he saw and frankly, so are we. This is the best-documented case on record, and still we have been unable, in spite of thorough investigation, to find the vehicle or other stimulus that scared Zamora to the point of panic.[26]

The air force closed Project Blue Book in mid-December, 1969. Of the more than 12,000 sightings it had recorded, 701 remained unidentified. Yet the air force declared that "no UFO reported, investigated and evaluated by the Air Force has ever given any indication

Congressman Acknowledges Seeing a UFO

Ohio representative Dennis Kucinich, a two-time presidential candidate, is perhaps the only member of Congress to admit having seen a UFO. In 2007, during a televised debate among the candidates, host Tim Russert asked Kucinich, "Shirley MacLaine writes in her new book that you sighted a UFO over her home in Washington state, that you found the encounter extremely moving, that it was a triangular craft, silent and hovering, that you 'felt a connection to your heart and heard directions in your mind.' Now, did you see a UFO?"

"Yes, I did," Kucinich responded. He added that former president Jimmy Carter had also seen a UFO and that more Americans had seen UFOs than approved of George W. Bush's presidency.

When the debate was over, reporters asked Kucinich when he saw the UFO. He said it had been 25 years ago. Then they wanted to know what it looked like. He refused to provide details. While some political commentators ridiculed Kucinich, his admission did not seem to bother his constituents. He was easily reelected in 2008 and again in 2010.

Quoted in *UFO-Blog.com,* Dennis Kucinich–Shirley MacLaine UFOs at Democratic Debate Follow-Up, November 16, 2007.

of a threat to our national security. . . . There has been no evidence indicating that sightings categorized as 'unidentified' are extraterrestrial vehicles."[27]

Project Blue Book ended the government's official involvement in UFOs. As author Rupert Matthews notes, "Documents have emerged since then by way of leaks and freedom of information requests that show the military is still interested in UFO reports and in civilian investigators."[28]

Sightings Continue

In any event, the attitude of the US government toward UFOs really did not make much difference. Just a few weeks after the demise of Project Blue Book, two cross-country skiers in Finland reported that they had come across a round platelike object hovering near the ground. It directed a beam of light onto the ground through the thick mist. Moments later a 3-foot-tall figure (1m) stood inside the beam, holding a box that emitted a pulsing yellow light. Seconds later the light beam and figure disappeared, followed by an explosion. The strange object was gone. The two men reported symptoms that resembled those of radiation poisoning, though they recovered.

> **DID YOU KNOW?**
>
> Many British and American military personnel say they saw something crash into England's Rendlesham Forest in December 1980. Some said they actually saw a UFO and the crash site had high radiation levels.

Six years later, in 1975, reports of a strange object in the sky over Tehran, Iran, resulted in the dispatching of a jet to investigate. As the jet neared the object, its electronic systems began malfunctioning. A second jet came to help out. The pilots saw an aircraft with many bright lights. At one point the pilots felt so threatened that they nearly fired defensive missiles, but another malfunction—as if the mystery object was monitoring their systems—prevented the launch. At that point the mysterious craft zoomed off at high speed. Because of the number of people deemed reliable who observed the encounter both visually and on radar, this incident is regarded as one of the strongest ever recorded.

The sighting spotlight shifted to Belgium late in 1989. Many people—including Belgian Air Force personnel, air traffic controllers, and police officers—reported seeing triangular-shaped aircraft over a period lasting for more than a year. The descriptions were surprisingly uniform. What was especially striking was the ability of the mysterious aircraft to perform aerial maneuvers seemingly beyond the capabilities of terrestrial airplanes. For example, the flights often exceeded the speed of sound, but no one ever heard a sonic boom. The Belgian military could never explain the phenomenon.

"The Belgium wave [of sightings] has obtained classic status in UFO lore," says UFO researcher Billy Booth. "With over 1,000 witnesses, confirmed radar sightings, plane radar lock-ins, and military confirmations, the fact that an unknown craft moved across the country of Belgium cannot be denied."[29]

The Belgian sightings were hardly the last word. Even today, hardly a month goes by without media reports from somewhere on the globe of yet another sighting.

Yet of all the thousands of sightings from ancient to modern times, the one with the most staying power and the firmest grip on the popular imagination happened within a few days of Kenneth Arnold's 1947 experience. For 24 hours the eyes of the world focused on a small town in New Mexico, and then, like an exploding firecracker, the brief, intense light faded almost instantly. It lay fallow and forgotten for more than 30 years.

CHAPTER THREE

A Small Town in New Mexico

Rancher William "Mac" Brazel worked a small property known as the Foster Ranch near Corona, New Mexico. On June 14, 1947, he was on a routine sweep of the property with his son. According to a newspaper article, "They came upon a large area of bright wreckage made up [of] rubber strips, tinfoil, a rather tough paper and sticks . . . the rubber was smoky gray in color. . . . Considerable Scotch tape and some tape with flowers printed upon it had been used in the construction. At the time Brazel was in a hurry to get his round made and he did not pay much attention to it."[30] Brazel had not seen or heard anything that could account for the debris, so he had no idea what its original size or shape might have been.

On July 4, more than two weeks later, Brazel returned to the area and scooped up some of the debris. The following day he learned of the June 24 Arnold sighting and its aftermath. Because living conditions on the ranch were primitive (no electricity, running water, telephone, or radio), Brazel had missed the initial excitement and the following daily reports of other saucer sightings. Now he wondered if what he had found might be connected in some way.

Two days later he went into Roswell, the nearest large town to the ranch. Roswell and its environs had a huge military presence. The Roswell Army Air Field (RAAF) was just outside of town. It housed the elite

509th Composite Group, whose B-29 bombers had dropped the atomic bombs on Hiroshima and Nagasaki less than two years earlier.

Reliable Witness

Brazel told Sheriff George Wilcox that he might have found the wreckage of a flying disk. Like Kenneth Arnold, Brazel had a reputation as a straight shooter. If he said something, it almost certainly was true. No one ever accused him of lying.

Wilcox called the Roswell airfield. Major Jesse Marcel, the base intelligence officer, and another officer drove back to the ranch with Brazel.

A sign marks the location of a supposed UFO crash in 1947 near Roswell, New Mexico. On June 14 of that year a local rancher happened upon the wreckage of a craft that he could not identify.

They spent some time in the debris field trying to reassemble what they found, though without success. The two officers returned to their base the following morning, July 8, with some of the debris. Events moved swiftly after that.

Early that afternoon, Marcel and his material flew to the Fort Worth Army Air Field in Texas, where it could be examined. The base commander was General Roger Ramey. The RAAF was also under his command.

While Marcel was still in Fort Worth, the Roswell base information officer, Walter Haut, issued a press release. It is not clear whether he acted on his own initiative or whether Marcel or someone else ordered him to. While no copy of the original press release remains, the *Roswell Daily Record*—which broke the story under the headline "RAAF Captures Flying Saucer on Ranch in Roswell Region"—likely ran it verbatim, as its editors would have been under deadline pressure. The opening paragraphs read,

> **DID YOU KNOW?**
> New Mexico lieutenant governor Joseph Montoya visited the Roswell Army Air Field on July 7, 1947. Witnesses say he went into a hangar that purportedly contained alien remains, and emerged badly shaken.

> Roswell, N.M.—The army air forces here today announced a flying disc had been found on a ranch near Roswell and is in army possession.
>
> The Intelligence office reports that it gained possession of the [disc] through the cooperation of a Roswell rancher and Sheriff George Wilson of Roswell.
>
> The disc landed on a ranch near Roswell sometime last week.[31]

At the same time, Haut's press release had gone out nationally. Predictably, it ignited a firestorm. Phone lines leading into Roswell were swamped with calls from newspaper reporters seeking more details.

While the story buzzed around the country, more was developing locally. It seemed as if everyone was seeing saucers. That was not surprising with all of the region's military facilities. If aliens were indeed watching Earth, it seemed likely that they might focus on the area where major weapons were being tested.

Deflating the Balloon

In mid-afternoon on July 8, a crush of reporters alerted by Haut's press release thronged the Fort Worth base. Rather than meeting them in the usual briefing room, Ramey took them to his office where some of the wreckage was on display. He explained that the wreckage was from a weather balloon, not a flying saucer. He had them take some pictures. Marcel was present and appeared in some of the photos but did not say anything.

> **DID YOU KNOW?**
>
> The 1994 TV movie *Roswell* was nominated for a Golden Globe as Best Mini-Series or Motion Picture Made for Television, but did not win.

The headline the following morning in the *Roswell Dispatch* read "Army Debunks Roswell Flying Disk As World Simmers With Excitement."[32] That afternoon, the *Daily Record*, the *Dispatch*'s competitor, began its own story by saying, "An examination by the army revealed last night that mysterious objects found on a lonely New Mexico ranch was a harmless high-altitude weather balloon—not a grounded flying disk. Excitement was high until Brig. Gen. Roger M. Ramey, commander of the Eighth air forces headquarters here cleared up the mystery."[33]

Across the page was another story, featuring Wilcox. It said that he had been cast "into the role of leading man in the world comedy which developed over the purported finding of a flying saucer."[34] Use of the word "comedy" suggests that the newspaper had bought into the official story.

For many Americans of that era—probably the vast majority—the word of a government official, in this case an air force general, was enough. The furor about flying saucers at Roswell instantly vanished from the national consciousness.

Brazel sounded a note of defiance in an interview that appeared the same day. "I am sure what I found was not any weather observation bal-

The Town of Roswell

Today the town of Roswell is synonymous with UFOs, but it was not always that way. The first settlers came to the Roswell area in 1865, though they soon left because of a lack of water. An enduring settlement began four years later under Van Smith and Aaron Wilburn. Smith named it Roswell, his father's first name.

A notable early resident was John Chisum, who built a ranch a few miles away. At one point it was the largest ranch in the United States, with more than 100,000 head of cattle. John Wayne played him in the 1970 movie *Chisum*, which provided a sympathetic portrayal of Chisum's role in the bloody and brutal Lincoln County War.

In the early 1930s American rocket pioneer Robert Goddard came to Roswell to conduct test flights of his rockets and spent more than a decade there. Just before American's entrance into World War II in 1941, the US Army Air Force built the Roswell Army Air Field, which was used to train thousands of fliers during the conflict. The town also housed nearly 5,000 German and Italian prisoners of war, who were used in public works construction projects.

Today Roswell has a population of about 50,000 and is the county seat of Chaves County. Notable people from Roswell include actress Demi Moore and singer John Denver.

loon. But if I find anything else besides a bomb they are going to have a hard time getting me to say anything about it."[35]

Fading Away

Brazel apparently never found anything else and faded from the historical record. So did Roswell. Life returned to normal.

During the waning years of Project Blue Book in 1966, the air force—stung by criticism by pro-UFO groups of its apparent indifference to sightings—contracted with the University of Colorado to produce an independent evaluation of the evidence for the existence of UFOs. The evaluation leaders asked the two most prominent UFO groups to submit a list of their most prominent sightings for review. Neither group mentioned Roswell.

A prominent and highly respected ufologist named Ted Bloecher published *Report on the UFO Wave of 1947* the following year. The book listed 853 sightings in late June and July of that year. Bloecher did not believe Roswell was worthy of inclusion, dismissing it by saying, "Through a series of clumsy blunders in public relations . . . the story got blown up out of all proportion."[36]

Back in the News

Roswell's return to renown began with a chance encounter in 1978. UFO researcher and author Stanton Friedman was in Baton Rouge, Louisiana, to interview several people about UFOs for a TV program. Friedman, a nuclear physicist who had worked with several major American companies, was one of a handful of professional scientists who believed in the reality of UFOs. During a break in the filming, the station manager said, "The guy you really ought to talk to is Jesse Marcel. He handled wreckage of one of those saucers you are interested in when he was in the military."[37]

By then, Marcel was dying of cancer. He told Friedman that he wanted to set the record straight. And the story he was now telling was very different from what he had said more than three decades earlier.

Much had changed in the intervening three-plus decades, especially in terms of Americans' attitude toward their government. The Vietnam War, conspiracy theories about the 1963 assassination of President John F. Kennedy, the 1972 Watergate break-in of the Democratic National Committee's headquarters—incidents such as these and many more had undermined the willingness of Americans to believe what their govern-

> **DID YOU KNOW?**
>
> According to some accounts, military officials kept Mac Brazel in custody for several days to ensure that he would cooperate with their version of events.

ment said. They were much more likely to believe stories of a cover-up now than they had been back in 1947. And a cover-up was exactly what Marcel was alleging. His charges quickly gained traction.

He said that the material he was photographed with in Ramey's office on July 8, 1947, was not the material he had recovered from the ranch. This new material was from a genuine weather balloon. Friedman eventually contacted scores of people who in varying degrees backed up Marcel's version.

Raising the Stakes

Something even more startling soon emerged. Friedman gave a lecture about eight months after meeting Marcel. When it was over, someone in the audience told him that a man named Grady Barnett claimed to have come across a crashed saucer, surrounded by several bodies of humanoids. The supposed crash site was near Socorro, about 150 miles (241km) northwest of Roswell. Friedman's informant was not sure when the incident had occurred but thought it was at about the same time as Brazel's discovery. He added that Barnett said a group of University of Pennsylvania archaeologists studying a nearby Native American site stumbled onto the site not long afterward. Then truckloads of soldiers drove up and ordered all of them to leave.

This was a major addition to the Roswell story. As scientist B.D. Gildenberg, who worked at Alamogordo during this period, notes, "The original Roswell story was very cut and dried—a sheep farmer found some material that looked like it could be from outer space. The Army was able to identify it, after having first created a PR fiasco with premature speculation about its nature. . . . None of the original impact site witnesses reported bodies."[38]

Barnett had died in 1969 and could not vouch for the story's authenticity. Other people said they could. Lurid tales of alien bodies and threats to witnesses to keep their mouths firmly shut soon emerged. Now a tantalizing possibility surfaced. Rather than bits and pieces of foil and sticks, a real flying saucer and dead crew members might have been found.

Whatever the case, the Roswell story soon sprouted an entirely new set of legs with this information. Glenn Dennis, who had worked as an apprentice in a Roswell funeral home in 1947, became one of the principal players. Dennis said he had been contacted by base officials at the time of the Brazel incident, asking if the funeral home had child-sized caskets.

General Roger Ramey (left), commander of the Roswell Army Air Field, identifies metallic fragments found by a New Mexico rancher as pieces of a downed weather balloon. Many Americans accepted this explanation, but others continued to believe that an alien craft had crashed on Earth.

A Strange Encounter

Later that day Dennis went to the base on an unrelated matter. He saw a truck with strange pieces of wreckage. Then he went into the hospital to look for a friend of his, a nurse. She emerged from an examining room, looking very frightened. "My gosh, get out of here or you're going to be in a lot of trouble,"[39] she blurted out.

Soon afterward he was accosted by several MPs (military police) who told him to keep quiet about everything he had seen or heard that day. Dennis protested that he was a civilian and could not be ordered around.

"Yes, we can," one replied. "Somebody will be picking your bones out of the sand."[40]

Dennis met his nurse friend the following day. She told him that she had helped do an autopsy on an alien body brought in under conditions of absolute secrecy along with two others. Each one was just under 4 feet (1.2m) in height and stank horribly. Their faces were not human, and the arms had four fingers which ended in suction cups. After the autopsy, the bodies had been packed up and flown away.

A few days later, Dennis learned that the nurse had been transferred to England. He wrote to her, but the letter was returned with the notation "Return to sender—deceased." When he was interviewed, Dennis withheld her name. Later he said it was Naomi Selff. No one by that name had worked at RAAF.

The Country Takes Notice

In the meantime, Friedman had been busy. Along with authors Charles Berlitz and William Moore, he interviewed many people. In 1980 they

> **DID YOU KNOW?**
> For many years, Walter Haut—who issued the famous press release about the Roswell UFO—denied any direct knowledge of aliens. However, in an affidavit released after his death in 2005, he said that he was taken to a hangar where he saw alien corpses.

published a book entitled *The Roswell Incident*. Roswell was back in business. After being ignored for more than three decades, the incident moved front and center among UFO researchers.

Roswell also moved into the realm of popular culture. It was featured in an episode of the TV series *Unsolved Mysteries* in 1989. Researchers Kevin Randle and Donald Schmitt published the book *UFO Crash at Roswell* in 1991. Friedman, now collaborating with Don Berliner, wrote *Crash at Corona: The U.S. Military Retrieval and Cover-Up of a UFO* the following year. (The Brazel site was actually closer to the town of Corona than to Roswell.) In what was becoming a case of dueling witnesses, Randle and Schmitt responded with *The Truth About the UFO Crash at Roswell* in 1994. The two camps seemed primarily divided by which sets of statements from their respective witnesses seemed more valid.

Thus the location of the alleged crashed saucer kept changing, according to additional witnesses who kept stepping forward. At one point it seemed to be more than 100 miles (161km) from Roswell. In another version, it was much closer. In one scenario, two saucers had collided. The dates of the crash also varied.

Under pressure from New Mexico congressman Steven Schiff, who felt that this increasing interest in the incident was being disregarded, in 1995 the air force released what they claimed was the official version of events called *The Roswell Report: Fact Versus Fiction in the New Mexico Desert*. Yes, there had been a cover-up at Roswell, the air force admitted. But it had nothing to do with UFOs. Instead, they were trying to protect a top-secret surveillance operation called Project Mogul, located at the Alamogordo base. In 1947 tensions with the Soviet Union had been particularly intense. The United States knew that the Soviets were working to develop an atomic bomb.

The Secret Gets Out

Project Mogul consisted of a series of high-altitude balloons specially designed to test the atmosphere to see if the Soviets had begun nuclear testing. One key element of the report cited the newspaper story in which Brazel had said the date of his finding was June 14. According to records, Mogul #4 had been launched on June 4 and soon afterward disappeared from radar. It seemed likely that this was the balloon that had crashed. No one could confuse it with a weather balloon.

The report did not satisfy everyone. For one thing, the report admitted that a great deal of documentation involving the RAAF dating to 1947 had disappeared. Critics also complained that some evidence had been ignored. The controversy over Roswell continued.

A Roswell resident reenacts the alien autopsy said to have taken place at the Roswell Army Air Field in 1947. A nurse who supposedly took part in the supersecret operation leaked word of it to a friend. Later efforts to locate the nurse proved fruitless.

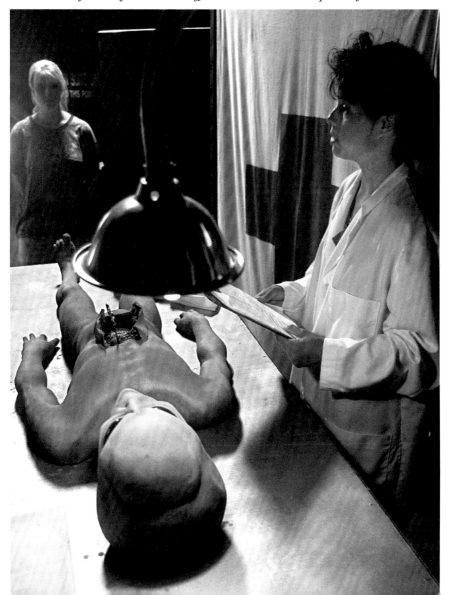

The Mysterious MJ-12

In 1984, documentary film producer Jaime Shandera—whose subjects had included UFOs—found a roll of camera film on his doorstep. After developing the film, he found he had in his possession documents that appeared to authorize formation of a secret group of high-ranking military officers, government officials, and prominent scientists whose purpose was to study UFOs. While the documents are almost certainly forgeries, there is a strong possibility that such a group, known as Majestic-12 or MJ-12, did exist—and may yet still exist.

Its purpose was to study the remains of UFOs and their occupants from Roswell and other locations and to keep a tight lid on this information. Projects Sign and Grudge, while publicly dismissive of UFOs, had been established to help MJ-12 in its work.

The reality of MJ-12 remains uncertain. Author Leslie Kean maintains that "it remains abundantly clear, despite the lack of official recognition, that a highly secret committee (perhaps one of several) was established in 1947 to deal with the greatest secret on Earth—a secret which remains classified to this day." On the other hand, Mark Pilkington, who has written extensively on UFOs, says, "The MJ-12 papers . . . represent [ufologists'] only hope of proving a government UFO conspiracy; even though some of those who accept that the documents are bogus still defend them as disinformation pointing to a deeper truth."

Leslie Kean, *Need to Know: UFOs, the Military, and Intelligence*. New York: Pegasus, 2007.

Mark Pilkington, *Mirage Men: An Adventure into Paranoia, Espionage, Psychological Warfare, and UFOs*. New York: Skyhorse, 2010.

Schiff was not the only politician to become involved. John Podesta, who served as chief of staff for President Bill Clinton and overseer of President Barack Obama's installation into the White House, said in 2002, "It is time for the government to declassify records that are more than twenty-five years old and to provide scientists with data that will assist in determining the real nature of this investigation."[41] Two years later, then–New Mexico governor Bill Richardson went even further:

"The mystery surrounding this crash has never been adequately explained. . . . It would help everyone if the U.S. Government disclosed everything it knows."[42]

Apparently those requests went unheeded. Richardson tried to secure the Democratic presidential nomination in the 2008 election. He said he would use the power of his office to investigate the situation if he were elected, but he dropped out of the race before the nominating convention. The authenticity of the supposed flying saucers remains uncertain.

The Reality of UFOs Lives On

But one thing is certain. Roswell is firmly on the map. In 1992 Dennis and Haut opened the Roswell International UFO Museum and Research Center. Four years later the museum and the Roswell Chamber of Commerce organized a UFO Festival that currently is held on the closest weekend to July 4 and attracts thousands of visitors each year. A number of hotels, motels, and businesses bear alien- and space-related names. Roswell street lights are shaped like the eyes of aliens. The town seems likely to remain a popular destination for those who believe in the reality of UFOs.

CHAPTER FOUR

UFO Encounters Get Personal

For all the attention that Roswell, Kenneth Arnold, and the myriad of other sightings in the late 1940s and all through the 1950s generated, no actual efforts by aliens to impose their will on earthlings had been noted. That was about to change.

Betty and Barney's Not-So-Excellent Adventure

About midnight on September 19, 1961, Betty and Barney Hill were driving to their home in Portsmouth, New Hampshire, after a brief Canadian vacation. They thought they would arrive home about 3:00 that morning.

Near the small town of North Woodstock, New Hampshire, Betty saw an intense white light that appeared to be following them. Barney stopped the car, got out, and peered at the object through binoculars. He saw a strange, wingless, flying object with several figures wearing uniforms gazing at him through a row of windows.

The sight terrified him. He leaped back into the car, gunned the engine, and fled. The Hills arrived home without further incident. One thing seemed odd, though. It was 5 a.m. They were not sure what had caused the delay.

"We entered our home, turned on the lights, and went over to the window and looked skyward," Betty wrote. "Then Barney said, 'This is the most amazing thing that has ever happened to me.' We both wondered if 'they' would come back."[43]

Odd Occurrences

The next day they noticed numerous highly polished markings about the size of fifty-cent pieces on their car. Betty placed a small compass next to them and the needle spun wildly. They also noticed that their watches had stopped working.

Betty and Barney Hill make use of a UFO diagram as they describe their abduction by aliens in the 1960s. The Hills' story is probably the best-known account of a UFO encounter that included direct contact with extraterrestrial beings.

The Hills notified the air force about what had happened. After some initial interest, the air force stopped further investigation. The official report said there was not enough data to proceed, even though the Hills had provided considerable detail.

By then Betty had begun having terrifying nightmares. Barney developed stomach ulcers. They could not account for the cause, except that it seemed to have something to do with the UFO sighting. They even drove back to the point where they had seen the alien craft to see whether that might jog their memories. It did not.

Finally, late in 1963, they began seeing psychologist Benjamin Simon. Simon put them under hypnosis to see whether he could determine what might have happened during those missing two hours.

What Really Happened

Though it required numerous sessions, eventually a very different story emerged. The Hills had not just sped away. Instead, Barney turned onto a side road without any conscious thought. Soon a group of humanoids about 5 feet (1.5m) tall with huge eyes, no noses, and slit-like mouths blocked the car. They yanked open the doors and escorted the Hills through the woods to their spacecraft.

As Betty recalled, "They are practically carrying him [Barney]. When he was being taken through the woods the tips of his shoes were dragging on the ground. . . . With me it was different. I was more or less walking under my own power."[44]

Once inside the spacecraft, Barney and Betty were taken to different rooms. The humanoids removed some of their clothing and subjected them to physical examinations. Barney's memory remained hazy. Betty, though, said that the creatures took skin and hair samples. "I sensed

> **DID YOU KNOW?**
>
> In 1985 author Whitley Streiber was among the first people to report being abducted from within their own homes. He has written extensively about his experiences in books such as *Communion.*

> ## Alien Abduction Insurance
> In 1987 Florida insurance salesman Mike St. Lawrence read Whitley Streiber's book *Communion*, which deals with alien abductions. When he finished, St. Lawrence checked his home owners' policy to see if it covered alien abductions. It did not. Sensing a need, he formed the UFO Abduction and Casualty Insurance Company and began offering a policy that promised a $10 million payout "in the event of the ABDUCTEE'S departure & return to planet Earth." The payout doubles under certain circumstances, such as the aliens insisting on conjugal visits with the person whom they had abducted. Unlike other insurance plans—which normally require ongoing payments—alien abduction insurance has just a one-time payment of $19.95. Thousands of people have purchased the policy, though St. Lawrence says that many name themselves as the beneficiary, frame the policy, and hang it on their home or office walls. He adds that he has actually paid some claims, which, as he explains in a television interview, require "the signature of an onboard authorized alien" to process. This policy has one catch: Payouts are made gradually, at a rate of $1 each year, rather than in a lump sum.

their excitement when they examined my skin through what I thought was some sort of microscope. . . . I assume they were examining my cell structure or genetic code."[45]

One part was particularly unpleasant. The creatures poked a long needle into her stomach, though the discomfort subsided when one of the humanoids, taller than the others and apparently the leader, leaned over her and began a conversation. At one point he showed her a map purporting to show where they were from.

Betty asked for something she could show other people to prove they had had this experience. The leader handed her a book but snatched it back at the insistence of the other humanoids. Then they took the Hills back to their car, at which point their memory returned. Later, Betty was able to recall the essential features of the map and make a drawing. While astronomers who processed it have not agreed, some believe it shows

Zeta Reticuli, a two-star system about 37 light years away. Each of the two stars appears to be similar to our sun and could have planets.

Keeping Mum

The Hills remained close-mouthed about the incident, and the details did not fully emerge until the publication in 1966 of *The Interrupted Journey* by UFO researcher John Fuller. As had been the case with Arnold, a number of other people reported similar experiences.

The story was the first indication that UFO sightings might have taken on a whole new dimension. It also made a "star" of Betty Hill, who often appeared at UFO gatherings until her death in 2004. Barney died in 1969 at the age of 46 from a cerebral hemorrhage. The Hills' saga was the beginning of the "Abduction Era," which would eventually overshadow "mere" UFO sightings and dominate much of the UFO dialogue up to the present day.

Encounter in the Night

What appears to be the first published attempt by purported aliens to kidnap earthlings took place the day after the first reports of the 1896 California airship sightings appeared in print. Two men, Colonel H.G. Shaw and Camille Spooner, were traveling in Shaw's horse and buggy when they encountered three strange figures on the road. The figures were about 7 feet (2m) tall and very thin. Shaw apparently was a man not easily frightened. He got out and tried to talk to them.

> **DID YOU KNOW?**
>
> Victims of alien abductions often describe their captors as having gray skin, huge heads, and large eyes.

Communication proved to be impossible. Shaw calmly watched and made extensive mental notes as the strange figures continued their conversation among themselves. "Their hands were quite small and delicate, and that their fingers were without nails," Shaw writes. "Their feet, however, were nearly twice as long as those of an ordinary man, though they were narrow, and the toes were also long and slender. I noticed, too, that they were able to use their feet and toes much the same as a monkey."[46]

He further noted that their skin was similar to velvet and very smooth. They had no facial hair, tiny ears, ivory-colored noses, and small, seemingly toothless mouths. "That and other things led me to believe that they neither ate nor drank, and that life was sustained by some sort of gas," Shaw notes. "Each of them had swung under the left arm a bag to which was attached a nozzle, and every little while one or the other would place the nozzle on his mouth, at which time I heard a sound of escaping gas."[47]

Attempted Abduction

While Shaw was observing them, the creatures were doing the same to him and Spooner, aided by intense, penetrating lights the size of a hen's egg. Then they moved into action:

> Finally they became tired of examining us and our horse and buggy, and then one of them, at a signal from one who appeared to be the leader, attempted to lift me, probably with the intention of carrying me away. Although I made not the slightest resistance he could not move me, and finally the three of them tried it without the slightest success. They appeared to have no muscular power outside of being able to move their own limbs.[48]

Failing in their mission to move him, they turned away and flashed their lights. To Shaw's astonishment, the illumination revealed an airship consistent with the descriptions of other witnesses during this period. It was about 150 feet (46m) long, with a maximum diameter of about 25 feet (7.6m) that tapered to points at each end. A door opened in the side and the visitors quickly disappeared within.

Once again Shaw showed no fear: "Just before it started I struck it with a rock and it gave no sound. It went through the air very rapidly and expanded and contracted with a muscular motion, and was soon out of sight."[49]

Fighting Back Against Would-Be Abductors

Longtime UFO researcher Ann Truffel interviewed more than 200 people who claimed to have been abducted by aliens between the 1960s and 1980s. To her surprise, dozens of them reported that they had managed to fend off their abductors. Truffel divided their responses into nine different techniques and published her results in 1988 in a book called *How to Defend Yourself Against Alien Abduction*.

Some, such as "Physical Struggle: Fight back" and "Repellents: Use time-tested fend-off substances" appear to be fairly obvious and would perhaps be effective in any threatening confrontation. Others, such as "Mental Struggle: Block their mind control" require more advance preparation. It requires sustained willpower, a sense that one's rights are being violated and a corresponding sense of outrage, and a conviction that the technique will work.

The book proved to be somewhat controversial in the UFO/alien abduction community. Some applauded Truffel's efforts. Others maintained that fighting alien abductors should be discouraged because of the opportunities for learning about them that the abduction offers. Still others said that the superior intelligence and technology of abductors make resistance futile.

Not all readers have taken her seriously. "It really worked!" notes an amazon.com reviewer. "I have not been abducted by aliens since I bought this book."

Quoted in Amazon.com, "It Really Worked!," review of *How to Defend Yourself Against Alien Abduction*, November 1, 2003. www.amazon.com.

Shaw quickly contacted his local newspaper, the *Stockton Evening Mail*, and his account appeared a few days later. Since his story cannot be independently verified, it remains a somewhat tantalizing footnote in UFO history.

Incident in Brazil

Though his story is nowhere near as well known as the Hills', Brazilian farmer Antonio Villas Boas had an experience that actually predated them by four years. In early October 1957, Boas and his brother saw a bright light in their farmyard as they were going to bed. It disappeared before they could get up to investigate. A week later they were plowing in the evening (to avoid the intense heat of the day) when something round and emitting a bright light appeared at the far end of the field. Boas tried to get closer, but it sped away.

The following night Boas was working by himself. A mysterious egg-shaped craft began hovering close to him, then came to rest nearby. He gunned his tractor and tried to flee, but the engine quickly died. He hopped off and started running. Moments later something grabbed his arm. It was a humanoid, about 5 feet (1.5m) tall. Boas knocked it down, but before he could flee three more—all larger than the first—pounced on him and carried him inside the spacecraft.

Boas claimed that it was only their superior numbers that overcame his resistance. "All seemed strong but not so strong that had I fought with one of them one at a time I should have been afraid of losing," he said. "I believe that in a free-for-all fight I could face any single one of them on an equal base."[50]

> **DID YOU KNOW?**
>
> Released in 1977, *Close Encounters of the Third Kind* was the first movie to show aliens as described by abduction victims.

An Unpleasant Examination

Once inside, Boas took note of their appearance. "All of them wore a very tight-fitting siren-suit, made of soft, thick, unevenly striped grey material," he said.

> This garment reached right up to their necks where it was joined to a kind of helmet made of a grey material that looked stiffer and was strengthened back at nose level. . . . Above their eyes, those helmets looked so tall that they corresponded to what the double of the size of a normal head should be. . . . Right on top, from the middle of their heads, there sprouted three round silvery metal tubes."[51]

They stripped him, took blood samples, rubbed him with a strange gel, then left him alone. They pumped a gas inside the room and he became violently ill.

Moments later a beautiful woman joined him. When she left, the aliens returned Boas's clothing and took him on a tour of the spacecraft. It was cut short when he tried to slip an unusual clocklike instrument into his pocket to verify his experience. He was promptly thrown out, the ship took off, and Boas was alone in his field. He suffered some strange illnesses afterward—headaches, nausea, vomiting, lesions on his skin—which were consistent with radiation poisoning, but they cleared up within a few weeks.

UFO researcher Joao Martins heard of Boas's experience and interviewed him. Unlike the Hills, Boas did not need hypnosis or any other prompt to recall what happened. The two men agreed to keep the story quiet so copycats would not infringe on its authenticity. It did not emerge for nearly a decade. Boas went on to become a lawyer and maintained the truth of his story until his death in 1992.

> **DID YOU KNOW?**
>
> In 1967 Nebraska police officer Herbert Schirmer said he saw a flying saucer but drove away. Later analysis showed that he could not account for 20 minutes of his patrol, and under hypnosis he said he had been taken aboard the spacecraft.

Logger Gets Cut Down

Another famous case involved a young logger named Travis Walton, who was working in an Arizona forest in 1975. As his work crew finished their work and headed home, they saw a disc-shaped object about 20 feet (6m) in diameter and 8 feet (2.4m) to 10 feet (3m) thick hovering motionless and silently over a nearby clearing.

"My God!" one of the men yelled. "*It's a flying saucer!*"[52]

The driver stopped the truck to get a better look. Walton, 22 years old at the time, got out and walked up to the object. It began rocking back and forth and emitting an increasingly louder noise. Before Walton could react, a beam of light knocked him down.

The rest of the men panicked and sped off. Several minutes later, they decided to go back and help Walton—or recover his body. When they returned to the site, they found no sign of either Walton or the mysterious object. The ground was undisturbed. When they contacted police, they came under suspicion of drug use. Then the possibility that they had murdered Walton emerged.

Five days later, Walton's brother-in-law Grant Neff received a phone call. It was Walton. "They brought me back!" he yelled. "I'm out here in Heber [a small town about 10 miles (16km) from the abduction site], please get somebody to come and get me!"[53] Though Neff suspected a prank, he rushed to retrieve Walton, who was confused, hungry, dehydrated, and in need of a shave.

The Horror

When he recovered, Walton told a bizarre story. After being jolted in the field, he had awakened, slightly dazed and in intense pain, on a table in a clean room that suggested a hospital. It did not take long for him to realize his mistake. "The sudden horror of what I saw rocked me as I realized that I was definitely not in a hospital. I was looking squarely into the face of a horrible creature! It looked steadily back at me with huge, luminous brown eyes the size of quarters. I looked frantically around me. There were three of them!"[54]

> **DID YOU KNOW?**
>
> Brazilian concert pianist Luli Oswald said she and the car she was driving were taken inside a UFO in 1979 and then examined and released.

The strange trio—about 5 feet (1.5m) tall, clad in orange jump suits and bald with high foreheads—moved toward him. When Walton leaped off the table and grabbed a nearby glass tube, they ran out of the room. Walton pursued them and soon found himself alone in a room, empty except for a large chair with buttons and dials on the armrests. As he fiddled with them, the room darkened.

The 1993 movie Fire in the Sky *depicted Travis Walton's 1975 UFO encounter in an Arizona forest. Walton wrote two books describing his abduction and the medical examination that he said was conducted by three aliens.*

The lights came back on, and a considerably more imposing human-like figure wearing a transparent helmet and a velour-like blue suit entered. This figure seemed well intentioned, so Walton followed him out of the UFO into what appeared to be a large hangar containing several similar aircraft. Moments later several creatures pounced on Walton, forced him onto a table, and put a mask over his face. Soon he lost consciousness.

He came to at the Heber gas station. A UFO, which Walton estimated at more than 40 feet (12m) in diameter, hovered nearby. Moments later it shot away at great speed. "The most striking thing about its departure was its quietness," Walton said. "It seemed impossible that something so large, moving through the atmosphere at such speed, would not have shrieked through the air, or even broken the sound barrier with a sonic boom. Yet it had been totally silent!"[55] Walton later learned that five days had elapsed. He thought he had been gone just a few hours.

In the next months, Walton and the other men passed several lie detector tests. When similar tests were administered again 20 years later, the results were the same. Walton's case remains as one of the few to have witnesses who corroborated significant portions of his account, though it remains somewhat controversial. He wrote a book about his experiences, *The Walton Experience*, in 1978. Fifteen years later Paramount Pictures released *Fire in the Sky*, a movie based on the book, though Walton criticized it because the events portrayed inside the UFO were different from his description. In 1997 he wrote a revised and updated version of his earlier book, entitled *The Walton Experience: Fire in the Sky*. He remains active in the UFO community, with frequent appearances at UFO conventions.

Playing the Numbers

These cases barely represent the tip of the abduction iceberg. According to a 2006 article in *Communication Quarterly*, "More and more Americans are warming to the idea that alien abductions could be real. Consistently, Gallup polls from the past 20 years have indicated that approximately half of those asked believe that UFOs are real."[56]

While not everyone agrees that UFO sightings and possible abductions are directly connected, the sense that "where there's smoke there's fire" seems to be growing. If UFOs are indeed prowling around Earth, it might make sense that they also want to get a close-up view of humanity.

CHAPTER FIVE

What Should We Believe?

Shortly after the curtain rises on William Shakespeare's most famous play, *Hamlet*, the title character sees his father's ghost. He turns to his close friend Horatio and says, "There are more things in heaven and earth, Horatio, than are dreamt of in your philosophy."[57]

This could be a good motto for anyone seeking the truth about UFOs. In the absence of one sighting that everyone agrees definitely took place, the argument rages back and forth—are we alone in the universe, or not? If we are not alone, have we been visited? To a large extent, the answer depends on one's point of view, or philosophy. Some people tend toward belief, while others remain skeptical.

Lunchtime Query

Skeptics note the "Fermi paradox." In 1950 noted physicist Enrico Fermi was having lunch with some fellow scientists. Fermi was one of the towering figures in twentieth-century science and a vital contributor to the success of the Manhattan Project, which developed the atomic bomb during World War II. The conversation turned to the subject of UFOs.

At one point, Fermi asked, "Where is everybody?"[58] In a universe that could be teeming with other civilizations—many of them millions of years older than ours—he was wondering why Earth had not been colonized. Or at least why aliens had not made their presence known.

To some people, the answer is obvious: Aliens *have* made their presence known. They point to massive structures such as the pyramids in Egypt, each of which contains more than 2 million stone blocks weighing at least 3,000 pounds (1,361kg) and leveled to within a fraction of an inch over a base that spans more than 13 acres (5.3ha). How could they have been built to such exacting measurements with the seemingly rudimentary technology of the time? Construction would also have been an enormous drain on manpower. According to most estimates, the project required tens of thousands of workers over a period of at least two decades in an agrarian economy that could ill afford to lose so many men for so long.

Despite solid archaeological evidence of the human origins of Egypt's pyramids, some UFO watchers attribute their construction to extraterrestrial beings. One author suggests that the massive stone blocks were lifted into place by an alien spacecraft.

J. Allen Hynek: Scientist and Believer in UFOs

J. Allen Hynek was the first professional scientist to publicly express support for the existence of UFOs even though, initially, he says, "I thought they were a lot of junk and nonsense." He became a convert during his involvement with the air force's Projects Sign, Grudge, and Blue Book.

Two things changed his mind. "One was the completely negative and unyielding attitude of the Air Force. They wouldn't give UFOs the chance of existing, even if they were flying up and down the street in broad daylight. . . . Secondly, the caliber of the witnesses began to trouble me. Quite a few instances were reported by military pilots, for example, and I knew them to be fairly well-trained, so this is when I first began to think that, well, maybe there [is] something to all this."

Hynek introduced a classification for UFO sightings called "close encounters." A close encounter of the first kind was seeing a UFO so closely that it could not be confused with anything else. The second kind involved physical effects the UFO produced, such as scorching. In the third kind, humanoids emerged from their spacecraft.

Hynek devoted his life to UFO research until his death in 1986. He had a distinguished career as an astronomer and teacher in addition to his UFO accomplishments.

Dennis Stacy, "Close Encounter with Dr. J. Allen Hynek: An Interview with the Dean, 1985," Computer UFO Network. www.cufon.org.

Help from Above?

According to author Erich von Däniken, extraterrestrials lent a hand—or more accurately, a spacecraft they employed to do the heavy lifting. Von Däniken expressed his views that Earth had been visited by aliens in his 1966 book *Chariots of the Gods*. He contended that the building of the pyramids was just one of many contributions aliens have made to human civilization. Though the scientific community scoffed at the book, it was very popular when it was published and still sells briskly.

That same year, noted astronomer Carl Sagan wrote in *Intelligent Life in the Universe*, "Our tiny corner of the universe may have been visited thousands of times in the past few billions of years. At least one of these visits may have occurred in ancient times."[59]

If he is right, these visits must have been considerably more frequent than the "at least one" he suggests. Though many possible sightings were recorded hundreds or even thousands of years ago, their persistence is striking. Unfortunately, the lack of detail in many of these accounts provides us with little or no concrete information about what they might have been observing.

Seeking Explanations

Many sightings can likely be explained on the basis of natural phenomena that are scientifically understandable today but were frighteningly unknown millennia ago. For example, some commentators have suggested that Thutmose's "fire circles" might have been meteor swarms. These result from the breakup of meteors and glow brightly as they enter the earth's atmosphere, though they would only have been visible at night. Since Thutmose's account does not say when the circles appeared—whether at night or during the day—no one knows for certain if meteor swarms could account for them.

> **DID YOU KNOW?**
>
> *Chariots of the Gods,* the 1966 book about alien visitors on Earth, has sold well over 60 million copies in more than 30 languages since it was published.

Other explanations turn inward. For example, the Berlin Wall dividing the German city of Berlin into east and west was dismantled on November 9, 1989. The Belgian UFO sightings began 20 days later, and some analysts suggest that anxiety over the likely reunification of Germany's eastern and western halves—which ultimately did take place—might have resulted in the sightings. Belgium had been invaded by Germany twice during the twentieth century—at the start of World War I and of World War II—and suffered brutal occupations by the Germans. According to this analysis, the prospect of a newly reunited and powerful

Germany might have created the right emotional environment for UFO sightings.

Whatever the merits of this argument, many people—especially psychiatrists and psychologists—maintain that anxiety and other mental states can account for sightings, abduction reports, and other UFO-related phenomena. Both British and American military authorities in the late 1940s believed that UFO reports peaked following periods of media publicity and could also be the result of Cold War fears. "Sightings may be psychological in origin,"[60] concluded a top-secret, high-level British group known as the Flying Saucer Working Party, convened during this period to study the phenomenon.

Deeper Fears

This view receives support from Carl Gustav Jung, regarded as one of the most important figures in classical psychiatry. He wrote *Flying Saucers: A Modern Myth of Things Seen in the Sky* in 1958. While remaining noncommittal about the actual existence of UFOs, Jung emphasized that accounts of UFO sightings could represent deep-seated fears that people buried in their subconscious minds.

> **DID YOU KNOW?**
>
> Carl Sagan convened a symposium to study the UFO phenomenon in 1969; it included scientific figures to argue both sides of the question.

Throughout history, people have looked to the heavens to ease their fears. In the past, people were comforted by worshipping their gods. The rise of science, according to Jung, weakened belief in gods and replaced it with confidence in technology. One of the twentieth century's best-known advertising slogans was the DuPont chemical company's slogan "Better Things for Better Living . . . Through Chemistry," which originated in 1935 and reflected this confidence. So, according to Jung, the gods of bygone eras morphed into technologically advanced visitors from extraterrestrial civilizations who might bring peace to Earth. He described this as a form of wish fulfillment that helped people cope with an increasingly perilous world.

At the time Jung wrote his book, Americans had undergone more than a decade of growing fear of the Soviet Union. This fear significantly increased in 1949, when the Soviets conducted their first atomic bomb test and the United States no longer was the only nation capable of inflicting nuclear destruction on its enemies. Almost immediately, American school children began "duck and cover drills" (in which they crawled under their desks and put their hands over their heads) to "protect" themselves in case of a Soviet airstrike. The government encouraged their parents to dig bomb shelters and stock them with food and water. The world, in short, was becoming increasingly dangerous.

This ongoing unease may have contributed to the rise of abduction stories in the 1960s and 1970s. At the time Jung wrote his book, people still appeared to regard aliens as benevolent, or at least nonthreatening. Now they were hostile, kidnapping people and performing often painful experiments on them against their will. Many felt compelled to record their accounts, often in lurid prose.

> **DID YOU KNOW?**
>
> Mysterious crop circles sometimes accompany UFO sightings. Often-intricate designs appear overnight in grain fields with no indication of who or what made them.

Questioning Abduction Accounts

Perhaps surprisingly, the famous abductee Betty Hill did not believe many of these abduction stories. "In this country, they say three million people have been abducted," she said. "Not once, but continuously. Do you realize that means three or four thousand people every night are being abducted? In this country alone? I don't know how the planes get through. . . . Actually, the real abductions, you've never heard of. Because nobody's writing any books about 'em."[61]

Author Paul Davies dismisses abduction stories from another perspective:

> Almost always these "ufonauts" were humanoid in form (sometimes dwarfs or giants), and often with descriptions that suggested something out of Hollywood central casting. . . . Another giveaway was the banality of the aliens' putative agenda, which seemed to consist of grubbing around in fields and meadows, chasing cows or aircraft or cars like bored teenagers, and abducting humans for Nazi-style experiments. Not what one would expect of cosmic superminds.[62]

The rise of abduction stories by hostile aliens may reflect what was happening in American society at large. The United States had undergone the disruption of the 1960s that began with the Cuban missile crisis in 1961—which nearly plunged the world into nuclear war—and continued with the assassinations of President John F. Kennedy and civil rights leader Martin Luther King Jr. The earlier view of technology as benign had been replaced by the idea that it could be harmful. Increasing pollution was one sign. Another came in the 1968 film *2001: A Space Odyssey*. The villain is an onboard computer named HAL, which tries to take over the spacecraft as it travels toward Jupiter. HAL's increasing creepiness reflects the unease with which people were now regarding machines that were supposed to make their lives easier and more comfortable. Above all, the Vietnam War and its aftermath had fractured the country, accompanied by the realization that government officials—both Republicans and Democrats—had consistently lied. Perhaps the lowest point of this pattern of dishonesty came in 1974 with the resignation of President Richard Nixon amid allegations that he was personally involved in the break-in of Democratic Party headquarters in the Watergate Hotel in Washington, DC.

Fitting into a Pattern

In this light, Jesse Marcel's allegation just a few years later that his superior officers had switched the material he had gathered with the help

The real-world dangers and unrest of the 1960s along with shifting attitudes toward technology might have influenced accounts of hostile aliens abducting humans. The 1968 movie 2001: A Space Odyssey, *about a computer that takes over a spacecraft, played into these fears.*

The British Look into UFOs

The US Air Force was not the only military organization investigating UFOs. In 1950, Henry Tizard—an eminent British scientist who is associated with the origins and development of radar—urged the formation of a panel to investigate flying saucer sightings that had occurred as far away as New Zealand. He did not want to dismiss them as mere delusions.

By that time the Soviet Union had developed nuclear weapons. Tizard, among others, thought it was important to make sure that UFOs were not Soviet aircraft carrying immense destructive power.

The result of this concern led to the formation of the Flying Saucer Working Party, or FSWP, in 1950. It consisted of five officers from several British intelligence agencies. While the government's official position was to downplay the sightings, the FSWP carried out its investigations under conditions of strict secrecy. It was not until 2001 that its report was declassified and opened to the general public.

After sifting through hundreds of reports in eight months, the panel felt only three were "reliable." Even those they eventually dismissed. They attributed all UFO sightings to mistaken identity, delusions, or outright hoaxes. The panel ended its report, "We recommend very strongly that no further investigation of reported mystery aerial phenomena be undertaken, unless and until some material evidence becomes available."

Flying Saucery Presents . . . The Real UFO Project, "Flying Saucer Working Party: Commentary." www.uk-ufo.org.

of Mac Brazel more than four decades ago, along with the testimony of other witnesses who said they had seen saucer crashes, did not seem surprising at all. They fit into the pattern of government cover-ups being revealed by the media on a regular basis and seemed to confirm the *Roswell Daily Record*'s suspicions several decades earlier: "The Army isn't telling its secrets yet," the paper complained. "But, SOMETHING has been found."[63]

British UFO researcher Rupert Matthews echoes the newspaper's sentiments: "There is no doubt at all that the USAAF launched a concerted and determined campaign to hide from the public the truth about what really happened in and around Roswell in early July 1947. The problem that researchers have been faced with is finding out what the truth was that was hidden so effectively."[64]

In the view of many people, the truth has been hidden so effectively that not even the date on which Brazel discovered the wreckage is certain. Brazel firmly fixed the date of his discovery as June 14 in a newspaper article published on July 9 after Ramey's all clear.

According to Brazel's son Bill, however, the discovery may have taken place considerably later, perhaps as late as July 4. Some suggest that the military put pressure on Brazel to name an earlier date to tie in more closely with Project Mogul. As Stanton Friedman notes, "The real, original, July 8 [a Monday] stories all said, 'found last week.'"[65]—which would be the period of July 1–7.

The problem of fixing exact dates—important as it is—pales in comparison with the question of whom to believe. For starters, the interviews with witnesses that reignited interest in Roswell did not begin until the late 1970s. Even the best-intentioned among them could be excused if their memories of events that happened more than three decades earlier might be a little hazy.

B.D. Gildenberg, who worked on the Mogul Project in 1947 and fully accepts the air force explanation, underscores this issue. He says that "Glenn Dennis told *Omni* magazine in 1995 that an earlier interview 'was the first time I tried to recall the whole experience in 40 years or more. . . . It's hard to get such old memories straight.'"[66] Another witness said to Gildenberg, "Since it's been 27 years . . . I may even have been influenced by other descriptions I've seen or heard in the interim."[67]

Seeing May Not Be Believing

The value of relying on eyewitnesses for an accurate account of events both past and present has also been called into question. An article in *Monitor on Psychology* dealing with eyewitness testimony in the criminal justice system asserts that "eyewitness identification often was not very good. Studies showed that witnesses often identified the wrong person from the photos (in one study, almost half the time) and that police interviewing techniques often hampered information gathering."[68] Most prosecutors take pains to bolster their cases with other forms of proof besides relying on witnesses.

While he was not an actual witness to the events at Roswell, air force general Nathan Twining quickly became involved in the situation. On July 8, Twining, the commander of the USAAF Air Material Command at Wright Field in Ohio, flew to Albuquerque, New Mexico, where he confirmed that the debris had nothing to do with any army air force experiments. Two months later, he sent a more comprehensive secret memo to the intelligence chief of the army air force. Parts of the memo have been cited in Roswell cover-up books as evidence of a crash: "(a) The phenomenon reported is something real and not visionary or fictitious. (b) There are objects probably approximately the shape of a disc, of such appreciable size as to appear to be large as man-made aircraft."[69]

> **DID YOU KNOW?**
>
> Animal mutilations are often cited as evidence of UFOs. In these incidents, parts of the animal are carved away and no footprints are left behind.

However, these books customarily omit another passage in Twining's memo that undercuts their argument: "Due consideration must be given [to] the lack of physical evidence which would undeniably prove the existence of these objects [UFOs]."[70]

The British Flying Saucer Working Party, formed in 1950, was blunt. The group concluded that "British Intelligence was informed by a member of the USAF investigation team that 'the . . . sensational report of the discovery of a crashed "flying saucer" full of the remains of very small beings, was ultimately admitted by its author to have been a complete fabrication.'"[71]

Interpreting a Photo

Both sides of the Roswell debate emphasize photos taken in 1947 of Marcel holding debris he had purportedly gathered as evidence of a cover-up. To Friedman, Marcel appears "embarrassed" because he knows he is perpetuating a lie. The material he is holding is something substituted for what he had originally found.

UFO researcher Philip J. Klass agrees that Marcel is covering up something. But it is not a lie. As reported on July 9, "Officers at the Roswell, N.M. air base received a stinging rebuke from the Army A.F. Headquarters in Washington, the United Press reported, for announcing that a 'flying disc' had been found on a New Mexico ranch."[72] In Klass's

Although many stories of UFO visits to Earth seem farfetched, the absence of definitive explanations for some accounts leaves open the possibility that alien spacecraft have made their way to Earth—or might do so at some point in the future.

view, Marcel's "embarrassment" reflects the "chewing-out" he has almost certainly just received from his superior officers.

Klass also agrees that the overall Roswell story represents a cover-up. But it has nothing to do with flying saucers. "In contrast to the flawed recollections of some Roswell witnesses, and the tall tales told by others, there is hard, credible evidence that shows that no ET craft crashed in New Mexico in 1947. This hard evidence shows that the only coverup is by those who accuse the Air Force and U.S. government of a crashed-saucer coverup."[73] In essence, Klass accuses TV producers and authors of books and magazine articles of keeping Roswell in the forefront of public consciousness simply for the purpose of making money.

Friedman—one of Klass's targets—fires back. "It is surprising to me that the recovery of crashed saucers in southeastern New Mexico in July 1947 is still a major bone of contention," he writes. "Despite all the research that had been done before the TV cameras came into play, the noisy negativists are still lying through their teeth about what actually happened and how the story came out."[74]

Ultimately Uncertain

The events at Roswell do not, on their own, prove the truth or falsity of UFO sightings. Neither do the numerous accounts of alien abductions, though Bruce Maccabee—who has researched many famous UFO sightings—notes that the story of Betty and Barney Hill "still stands as a strong reminder that we can't explain everything that happens—UFOwise or otherwise—and we'd better be prepared for further surprises."[75]

Skeptics do not deny that surprises can happen. Yet as Davies argues, "whatever lies behind that stubborn residue of hard-to-explain cases, I see no reason to attribute them to the activities of alien beings visiting our planet in flying saucers. UFO stories, like ghost stories, are fun to read, but cannot be taken seriously as evidence for extraterrestrial beings."[76]

Information abounds in today's world; answers can be found to most any question that one might have. Yet when it comes to UFOs, a definitive answer does not exist. The arguments on both sides can be very persuasive. The bottom line is this: After more than half a century of intense interest and study, we still do not know whether UFOs are real.

SOURCE NOTES

Introduction: Strange and Otherworldly Objects
1. Quoted in *Athena Review*, "First Voyage of Columbus: Meeting the Islanders (1492)," vol. 1., no. 3.

Chapter One: The Birth of Flying Saucers
2. Project 1947, "Fred Johnson Letter." www.project1947.com.
3. Quoted in *Oregon Journal*, "Carpenter Reports 'Discs' in Midwest," June 26, 1947. www.project1947.com.
4. Quoted in *Oregon Journal*, "Flying Disk Mystery Grows," June 26, 1947. www.project1947.com.
5. Kenneth Arnold, "Confidential," Project 1947. www.project1947.com.
6. Quoted in Bruce Maccabbee, "June 24, 1947: How It All Began—The Story of the Arnold Sighting." http://greyfalcon.us/Amerika.htm.
7. Quoted in Maccabbee, "June 24, 1947."
8. Quoted in *Boise (ID) Statesman*, "Harassed Saucer-Sighter Would Like to Escape Fuss," June 27, 1947. www.project1947.com.
9. Quoted in *Boise (ID) Statesman*, "Harassed Saucer-Sighter Would Like to Escape Fuss."
10. Quoted in Loren E. Gross, "UAL Flight 105—July 4, 1947," in *UFOs: A History, 1947*, pp.14–15. www.project1947.com.
11. Quoted in Curtis Fuller, "The Flying Saucers—Fact or Fiction?," *Flying*, July 1950. www.project1947.com.
12. *Portland Oregonian*, "'Flying Disc' Reports Come from Hundreds, in 28 States," July 6, 1947. www.project1947.com.
13. Kenneth Arnold, telegram. www.project1947.com.
14. Frank M. Brown, memorandum to the officer in charge, July 16, 1947. www.project1947.com.
15. Brown, memorandum.

Chapter Two: A Long History of Sightings

16. Quoted in Joseph Trainor, ed., *UFO Roundup*, vol. 1, no. 5, March 18, 1996. www.ufoinfo.com.
17. Quoted in Trainor, *UFO Roundup*.
18. *Washburn College Bible: King James Text, Modern Phrased Version*. New York: Oxford University Press, 1979, p. 1095.
19. Josef F. Blumrich, "The Spaceships of the Prophet Ezekiel," Unexplained & Paranormal Activities. www.bibliotecapleyades.net.
20. Quoted in Alien-UFOs.com, "UFOs in Ancient Times?," www.alien-ufos.com.
21. Quoted in *Jargon File*, "foo." www.catb.org.
22. *Time*, "Science: Foo-Fighter," January 15, 1945. www.time.com.
23. Quoted in Bill Yenne, *U.F.O.: Evaluating the Evidence*. New York: Smithmark, 1997, p. 45.
24. Mark Pilkington, *Mirage Men: An Adventure into Paranoia, Espionage, Psychological Warfare, and UFOs*. New York: Skyhorse, 2010, pp. 65–66.
25. Yenne, *U.F.O.: Evaluating the Evidence*, p. 74.
26. Quoted in National Investigations Committee on Aerial Phenomena, "Lonnie Zamora/Socorro NM Case Directory." www.nicap.org.
27. Quoted in Rupert Matthews, *UFOs: A History of Alien Activity from Sightings to Abductions to Global Threat*. New York: Chartwell, 2009, p. 178.
28. Matthews, *UFOs: A History of Alien Activity*, p. 179.
29. Billy Booth, "1989—the Belgium UFO Wave," About.com UFOs/Aliens. http://ufos.about.com.

Chapter Three: A Small Town in New Mexico

30. Quoted in Philip J. Klass, *The Real Roswell Crashed-Saucer Coverup*. Amherst, NY: Prometheus, p. 38.
31. Quoted in Rupert Matthews, *Roswell: Uncovering the Secrets of Area 51 and the Fatal UFO Crash*. New York: Chartwell, 2009, p. 21.
32. Dennis Balthaser, "The Other Roswell Newspaper," www.truthseekeratroswell.com.
33. Quoted in Matthews, *Roswell*, p. 35.

34. Quoted in Matthews, *Roswell*, p. 34.
35. Quoted in Klass, *The Real Roswell Crashed-Saucer Coverup*, pp. 21–22.
36. Quoted in Klass, *The Real Roswell Crashed-Saucer Coverup*, p. 22.
37. Stanton T. Friedman, *Flying Saucers and Science: A Scientist Investigates the Mysteries of UFOs*. Franklin Lakes, NJ: New Page, 2008, p. 218.
38. B.D. Gildenberg, "A Roswell Requiem," *Skeptic*, 2003. www.highbeam.com.
39. Quoted in Matthews, *Roswell*, p. 97.
40. Quoted in Matthews, *Roswell*, p. 95.
41. Quoted in Pilkington, *Mirage Men*, p. 300.
42. Quoted in Pilkington, *Mirage Men*, p. 301.

Chapter Four: UFO Encounters Get Personal

43. Quoted in Stanton Friedman and Kathleen Marden, *Captured! The Betty and Barney Hill UFO Experience*. Franklin Lakes, NJ: New Page, p. 33.
44. Quoted in Friedman and Marden, *Captured!*, p. 117.
45. Quoted in Friedman and Marden, *Captured!*, pp. 141–42.
46. Quoted in Alien-UFOs.com, "Colonel H.G. Shaw's Report of 1896," reported by the *Stockton (CA) Evening Mail*, November 25th, 1896. www.alien-ufos.com.
47. Quoted in Alien-UFOs.com, "Colonel H.G. Shaw's Report."
48. Quoted in Alien-UFOs.com, "Colonel H.G. Shaw's Report."
49. Quoted in Alien-UFOs.com, "Colonel H.G. Shaw's Report."
50. Quoted in Michael Naisbitt, "Interstellar Intercourse—the Abduction of Antonio Villas Boas," *UFO Digest*, May 8, 2007. www.ufodigest.com.
51. Quoted in Terry Melanson, "Antonio Villas Boas: Abduction Episode Ground Zero," Illuminati: Conspiracy Archives, May 10, 2005. www.conspiracyarchive.com.
52. Quoted in Travis Walton, "An Ordinary Day," in *Fire in the Sky*, third ed. New York: Marlowe, 1997. www.travis-walton.com.
53. Travis Walton, "Return," in *Fire in the Sky*.
54. Travis Walton, "The Aliens," in *Fire in the Sky*.

55. Walton, "Return."
56. Stephanie Kelley-Romano, "Mythmaking in Alien Abduction Narratives," *Communication Quarterly*, August 1, 2006. http://business.highbeam.com.

Chapter Five: What Should We Believe?

57. William Shakespeare, *Hamlet, Prince of Denmark*, act 1, scene 5, lines 167–68.
58. Quoted in Paul Davies, *The Eerie Silence: Renewing Our Search for Alien Intelligence*. Boston: Houghton Mifflin Harcourt, 2010, p. 116.
59. Quoted in *Encyclopedia of the Unusual and Unexplained*, "UFOs in Ancient Times." www.unexplainedstuff.com.
60. Flying Saucery Presents . . . The Real UFO Project, "Flying Saucer Working Party: Commentary." www.uk-ufo.org.
61. Quoted in Peter Brookesmith, "Abductee Betty Hill Slams Abduction Research," UFO Digest. www.ufodigest.com.
62. Davies, *The Eerie Silence*, p. 21.
63. Quoted in Matthews, *Roswell*, p. 35.
64. Matthews, *Roswell*, p. 118.
65. Friedman, *Flying Saucers and Science*, p. 224.
66. Gildenberg, "Roswell Requiem."
67. Gildenberg, "Roswell Requiem."
68. Kathryn Foxhall, "Suddenly, a Big Impact on Criminal Justice." *Monitor on Psychology*, July 2000. www.apa.org.
69. Quoted in Klass, *The Real Roswell Crashed-Saucer Coverup*, p. 28.
70. Quoted in Klass, *The Real Roswell Crashed-Saucer Coverup*, p. 29.
71. Flying Saucery Presents . . . The Real UFO Project, "Flying Saucer Working Party: Commentary."
72. Klass, *The Real Roswell Crashed-Saucer Coverup*, p. 22.
73. Klass, *The Real Roswell Crashed-Saucer Coverup*, p. 204.
74. Friedman, *Flying Saucers and Science*, pp. 217, 223.
75. Quoted in Friedman and Marden, *Captured!*, p. 14.
76. Davies, *Eerie Silence*, p. 22.

FOR FURTHER EXPLORATION

Books

Paul Davies, *The Eerie Silence: Renewing Our Search for Alien Intelligence.* Boston: Houghton Mifflin Harcourt, 2010.

Stanton T. Friedman, *Flying Saucers and Science: A Scientist Investigates the Mysteries of UFOs.* Franklin Lakes, NJ: New Page, 2008.

Stanton T. Friedman and Kathleen Marden. *Captured! The Betty and Barney Hill UFO Experience.* Franklin Lakes, NJ: New Page, 2007.

Leslie Kean, *Need to Know: UFOs, the Military, and Intelligence.* New York: Pegasus, 2007.

Philip J. Klass, *The Real Roswell Crashed-Saucer Coverup.* Amherst, NY: Prometheus, 1997.

Rupert Matthews, *Roswell: Uncovering the Secrets of Area 51 and the Fatal UFO Crash.* New York: Chartwell, 2009.

Rupert Matthews, *UFOs: A History of Alien Activity from Sightings to Abductions to Global Threat.* New York: Chartwell, 2009.

Mark Pilkington, *Mirage Men: An Adventure into Paranoia, Espionage, Psychological Warfare, and UFOs.* New York: Skyhorse, 2010.

Jacques Vallee and Chris Aubeck, *Wonders in the Sky: Unexplained Aerial Objects from Antiquity to Modern Times.* New York: Jeremy P. Tarcher/Penguin, 2009.

Bill Yenne, *U.F.O.: Evaluating the Evidence.* New York: Smithmark, 1997.

Internet Sources

Alien-UFOs.com, "Colonel H.G. Shaw's report of 1896," reported by the *Stockton (CA) Evening Mail*, November 25, 1896. www.alien-ufos.com/ufo-alien-discussions/15177-colonel-h-g-shaws-report-1896-a.html.

B.D. Gildenberg, "A Roswell Requiem," *Skeptic*, 2003. www.highbeam.com/doc/1G1-101495912.html.

Kal K. Korff, "What Really Happened at Roswell," *Skeptical Enquirer*, July/August 1997. www.csicop.org/si/show/what_really_happened_at_roswell.

Dennis Stacy, "Close Encounter with Dr. J. Allen Hynek: An Interview with the Dean, 1985," Computer UFO Network. http://www.cufon.org/cufon/hynekint.htm.

Websites

About.com UFOs/Aliens (http://ufos.about.com). A website moderated by longtime researcher Billy Booth; contains news, photos, videos, current and classic cases, and more.

Mutual UFO Network (MUFON) (www.mufon.com). Official website of MUFON; contains news, case files, merchandise, and a list of the organization's local chapters.

Project 1947 (www.project1947.com). Website of Project 1947, an ongoing research effort into documenting the start of the modern UFO era; features many original documents from participants and observers.

Search for Extra-Territestrial Intellingence Institute (SETI) (www.seti.org). Official website of SETI, whose mission is "to explore, understand, and explain the origin, nature and prevalence of life in the universe."

Travis Walton (www.travis-walton.com). Website of alien abductee Travis Walton, with book excerpts, updated event appearances, and more.

INDEX

Note: Boldface page numbers indicate illustrations.

Air Force, US
   attitude toward UFO reports, 60
   conclusions on Project Blue Book, 30–31
   report on Roswell incident, 42–43
Air Technical Intelligence Center, 26, 28
Alexander the Great, 22–23
alien abduction insurance, 49
alien abductions
   of Antonio Villas Boas, 52–54
   of Betty and Barney Hill, **47**, 48–50
   defending against, 52
   reasons for rise of reports in 1960s and 1970s, 63–64
   of Travis Walton, 54–55, 57
alien(s)
   autopsy of, 41, **43**
   descriptions of, 48, 50–51, 53, 55
animal mutilations, 68
Annunciation with St. Emidius, The (Crivelli), 23
Arnold, Kenneth, 8–10, **11**, 11–14, 15–16, 32
   conclusions of Air Force report on, 17

Baptism of Christ, The (Gelder), 23
Barnett, Grady, 39
Baxter, Al, 12
Belgium, 1989 sighting in, 32
Bequette, Bill, 13
Berliner, Don, 42
Berlitz, Charles, 41–42
Bethlehem, Star of, 27
Bible, 22, 27
Bloecher, Ted, 38

Blue Book. *See* Project Blue Book
Blumrich, Josef F., 22
Boas, Antonio Villas, 52–54
Booth, Billy, 32
Brazel, William "Mac," 33, 34, 36–37, 38, 66
Brown, Frank M., 16–17, 19
Buck Rogers, 6–7
Burroughs, Edgar Rice, 6

California, 1896 sighting in, 24
Carter, Jimmy, 30
Cassius, Dio, 23
Chariots of the Gods (Däniken), 60, 61
Chisum, John, 37
Cicero, 23
close encounters, classification of, 60
Close Encounters of the Third Kind (1977 film), 29
Columbus, Christopher, 4–5
Communication Quarterly (journal), 57
Communion (Streiber), 49
Crash at Corona: The U.S. Military Retrieval and Cover-Up of a UFO (Friedman and Berliner), 42
Crivelli, Carlo, 23

Dahl, Harold, 15, 17–19
Däniken, Erich von, 60
Davidson, William, 16, 19
Davies, Paul, 64, 70
Dennis, Glenn, 40–41, 45, 67
discus, 26

Egypt, ancient
   pyramids of, **59**, 59–60
   sighting in, 20–22
England, medieval, sightings in, 24

Ezekiel (biblical prophet), 22

Fermi, Enrico, 58
Finland, 1970 sighting in, 32
Fire in the Sky (1993 film), **56,** 57
First Men in the Moon. The (Wells), 6
Flash Gordon, 6–7
flying saucers, beginning of era of, 13, **18**
Flying Saucers (Jung), 62–63
Flying Saucer Working Party (FSWP), 62, 66, 68
foo fighters, 24–25, **28**
Friedman, Stanton, 38, 39, 41–42, 67, 69
From the Earth to the Moon (Verne), 6
Fuller, John, 50

Gelder, Aert de, 23
Gildenberg, B.D., 39, 67
Goddard, Robert, 37
Grudge. *See* Project Grudge

Haut, Walter, 35, 36, 41, 45
Hill, Barney, 46–50, **47**
Hill, Betty, 46–50, **47, **63
How to Defend Yourself Against Alien Abduction (Truffel), 52
Hynek, J. Allen, 27, 29, 60

Intelligent Life in the Universe (Sagan), 61
Interrupted Journey, The (Fuller), 50
Iran, 1975 sighting in, 31

Johnson, Fred, 10
Jung, Carl Gustav, 62–63

Kennedy, John F., 38, 64
King, Martin Luther, Jr., 64
Klass, Philip J., 69–70
Kucinich, Dennis, 30

Livy, 23

Maccabee, Bruce, 70

Madonna col Bambino e San Giovannino (painting), 23
Mahabharata (ancient Indian epic), 24
Manhattan Project, 58
Marcel, Jesse, 34, 38, 39, 64, 68–70
Martins, Joao, 54
Matthews, Rupert, 31, 66–67
McHenry, Joseph C., 16
medieval art, UFOs in, 23
Mogul. *See* Project Mogul
Monitor on Psychology (journal), 67
Montoya, Joseph, 35
Moore, William, 41–42

Nixon, Richard, 64
Nuremberg Chronicle (1493 illustrated world history), 23

Obsequens, Julius, 23
Omni (magazine), 67
opinion poll, on belief in UFOs, 57
Oswald, Luli, 55

Pilkington, Mark, 27
Pliny the Elder, 23
Podesta, John, 44
Portland Oregonian (newspaper), 15
Prajnaparamita Sutra (Sanskrit text), 24
Project Blue Book, 28, 29, 30, 37
    conclusions of Air Force on, 30–31
Project Grudge, 27–28, 44
Project Mogul, 42–43, 67
Project Sight, 26–27
pyramids, **59,** 60

Quantanilla, Hector, 29–30

Ramayana (ancient Indian epic), 24
Ramey, Roger M., 35, 36, **40**
Randle, Kevin, 42
Renaissance art, UFOs in, 23
Report on the UFO Wave of 1947 (Bloecher), 38
Richardson, Bill, 44–45
Roswell (1994 TV film), 36

Roswell, New Mexico, 33, **34**, 37
   as destination for UFO enthusiasts, 45
   UFO sighting in
      Air Force report on, **40**, 42, 33–35
      cover-up of, **43,** 68–70
Roswell Daily Record (newspaper), 35, 66
Roswell Incident, The (Berlitz and Moore), 41–42
Roswell Report, The (US Air Force), 42–43
Ruppelt, Edward, 28–29

Sagan, Carl, 61, 62
Schiff, Steven, 42, 44
Schirmer, Herbert, 54
Schmitt, Donald, 42
Shaw, H.G., 50–52
Sight. *See* Project Sight
sightings
   alternative explanations for, 61–63
   in ancient/medieval times, 20–24
   in Belgium (1989), 32
   in California (1896), 24
   classification of, 60
   in Finland (1970), 31
   in Iran (1975), 31
   of 1947, 9–10, 16, 17–18
      report on, 38
      in Roswell, 33–35
   questions about eyewitness testimony, 67–68
   during World War II, 25–26
Simon, Benjamin, 48
Skiff, Nolan, 13
Smith, Emil J., **11,** 14
Smith, Van, 37
Spooner, Camille, 50
St. Lawrence, Mike, 49
Star of Bethlehem, 27
Stevens, Ralph, **11,** 14–15
Stockton Evening Mail (newspaper), 52
Streiber, Whitley, 48, 49
survey, on belief in UFOs, 57

Taino Indians, 4–5
Thutmose III (Egyptian pharaoh), 20, **21,** 61
Time (magazine), 25
Tizard, Henry, 66
Truffel, Ann, 52
Truth About the UFO Crash at Roswell, The (Randle and Schmitt, 1994), 42
Twining, Nathan, 68
2001: A Space Odyssey (1968 film), 64, **65**

UFO Crash at Roswell (Randle and Schmitt, 1991), 42
UFOs (unidentified flying objects), **5, 69**
   British investigations into, 62, 66, 68
   continuing uncertainty about, 70
   descriptions of, **18, 28,** 51, 54
   in medieval/Renaissance art, 23
   origin of term, 29
   prevalence of belief in, 57
   See also sightings
Unsolved Mysteries (TV series), 42

Vandenberg, Hoyt, 27
Verne, Jules, 6

Walton, Travis, 54, 56, 57
Walton Experience, The (Walton, 1978), 57
Walton Experience: Fire in the Sky, The (Walton, 1997), 57
War of the Worlds (1938 radio broadcast), 6, 7, 12
Welles, Orson, 7, 12
Wells, H.G., 6
Wilburn, Aaron, 37
Wilcox, George, 34, 36

Yenne, Bill, 29

Zamora, Lonnie, 29–30
zeppelin airships, 24, **25**
Zeta Reticuli (star system), 50

PICTURE CREDITS

Cover: iStockphoto.com and Photos.com
© Mike Agliolo/Corbis: 5
AP Images: 34
© The Art Archive/Corbis: 21
© Bettmann/Corbis: 11, 40, 47
© Stefano Bianchetti/Corbis: 25
© Claudius/Corbis: 69
Landov/Reuters: 43
Gregory Macnicol/Science Photo Library: 18
Photofest: 56, 65
Thinkstock/iStockphoto: 59
Victor Habbick Visions/Science Photo Library: 28

ABOUT THE AUTHOR

Jim Whiting has written more than 115 nonfiction books for young readers and edited well over 150 more during his diverse writing career. He published *Northwest Runner* magazine for more than 17 years. He has also been an adviser to a national award-winning high school newspaper, a sports editor for the *Bainbridge Island Review*, and a writer and photographer for America Online. His articles have appeared in dozens of newspapers and magazines.